A THEORY OF COSMOLOGY

The Small Bang Theory
and
An Introduction to
The New Physics

A THEORY OF COSMOLOGY

The Small Bang Theory and An Introduction to The New Physics

by

DAVID FISKE

VANTAGE PRESS
New York Washington Atlanta Hollywood

FIRST EDITION

Published by Vantage Press, Inc.
516 West 34th Street, New York, New York 10001

Printed in the United States of America
Standard Book Number 533-03981-9

To my wife Ruth,
without whose patience and understanding this work could never have been done.

CONTENTS

List of Figures

ACKNOWLEDGMENT: Figures 1, 2, 3, 6, 7, and 8 were done by Sidney Witz.

Preface

This essay is not directed at the *faint of heart,* who might cringe from any intrusion into traditional principles. The exposition is concerned with entry into unorthodox physics. As such, the immediate reaction of some might be skepticism. However, if you, the reader, have an open mind and are willing to exercise it, the plea is made that you proceed. The following summary indicates the areas of development:

1. The *small bang* theory of cosmology is presented, whereby *something comes from nothing,* by virtue of the acceptance of negative mass.
2. The *emag hypothesis* is presented, whereby the heretofore "fictitious ether" is replaced by an ether of reality, namely, a *sea of emags.*
3. The emag is presented as the ultimate particle of matter, with its symmetrical component of negative mass, called the *other emag.*
4. The dimensions of the emag are presented, as compared to the dimensions of recognized subatomic particles, i.e., electron and proton.
5. The nature of the "photon" is presented as a wave of emags of variable number and mass.
6. The "photon," as a wave of emags or emag aggregates, is presented in a manner whereby the wave phenomena of light are explained.
7. The emag hypothesis is presented to explain the particulate nature of electromagnetic radiation.
8. The *emission* nature of electromagnetic radiation is presented as a consequence of the emag hypothesis.
9. The concept of "action at a distance" is brought into question, and replaced by the *direct action* of particles, which results directly from the emag hypothesis.
10. The long search for a unified "field" concept is abandoned in favor of the unified emag hypothesis, whereby all electromagnetic, coulomb, and gravitational forces are considered as emag phenomena.
11. The "stationary state" orbit of Bohr is evaluated in terms of the

emag hypothesis, and both the orbital frequency of the hydrogen electron and its orbital radius are determined in metric units.
12. The mass of the "photon" is determined, as based on the emag hypothesis.
13. The orbital velocity of the hydrogen electron is determined, as based on the emag hypothesis.
14. The second postulate of Einstein is brought into question and arguments are presented to show the inadequacy of this postulate.
15. The reality of the neutron as a discrete particle is challenged and abandoned.
16. A summary of electromagnetic forces and phenomena is presented.
    a. Distant effects are equated with gravity.
    b. Intermediate effects are equated with magnetism.
    c. Near effects are equated with electricity.
    d. Very near effects are equated with intranuclear force.

It would be the ultimate of *chutzpah* to presume that all statements made in the following exposition are correct. It is a large share of *chutzpah* to entertain an all-inclusive hypothesis, as presented. Time will tell how much of this presentation has merit. It is merely a first step. The author knows full well that confidence is no substitute for correctness.

I should like to acknowledge, with gratitude, Dr. John Marburger, then Chairman of the Department of Physics at the University of Southern California, who, for a period of time, guided me, then directed me, then pushed me and encouraged me to follow my inclinations.

I should also like to acknowledge Dr. Eugene Wong, Professor of Physics at the University of California in Los Angeles, for his patience and assistance in getting me over some troublesome hurdles.

A THEORY OF COSMOLOGY

**The Small Bang Theory
and
An Introduction to
The New Physics**

1. *The Small Bang Theory*

As we mature and expand our knowledge, which generally is based upon the experiences and teachings of others, we become slaves to tradition. Dogmas which have, apparently, withstood the test of time are imbibed as presented. Upon such foundations do we build. From time to time, cynics and doubters appear on the scene who ask, "Is such and such really the case?" or "What is the origin of this 'truth'?" Such doubters, frequently, are intimidated by the "weight of authority," and die on the vine. At other times, such doubters pursue their doubts, for whatever reason. The results of these exercises may not be futility. As examples, the tradition of Ptolemy gave way to Copernicus. The constant, orbital velocity of Copernicus gave way to Kepler's probings, and so on.

Since many of us *are* slaves to tradition, we frequently rear up in antagonism to the intrusion of a new doctrine which may shake the structure of tradition. Some of us are able to surmount this antagonism and to listen.

Cosmology is a matter for conjecture. This clearly is the case, since proof can never be obtained about the beginning of the universe. The best we can do is to determine whether or not reality is consistent with a theory of cosmology. It is apparent that it is possible that more than one theory can explain empirical reality. Ptolemy, with his circles and epicircles, described, quite satisfactorily, the motions of the planets with relation to the earth and the sun. Even Kepler was reluctant to abandon the circle and epicircle concept. Copernicus never did.

Similarly, arithmetic computation can give comparable results with a numerical system other than base 10. Does this mean that both systems are correct? In the case of the arithmetic example, the answer is yes. In the case of the solar planetary system, the answer is no.

If we should have the hubris to question the big bang theory, for example, or the theory of relativity, can we suggest viable alternatives? The answer is yes. And this is the basis of the following exposition.

Cosmology was a natural outgrowth of inquiry. How did *anything* begin? "In the beginning, God created the heaven and the earth." So says the Bible. And mankind has generally adopted and accepted the concept of some supernatural force which *created.* Perhaps the best example of this philosophy is the Christian dogma of *First Cause* or *Uncaused Cause.*

It was necessary for man to invent the idea of uncaused cause because rational man had learned, from experience, that every effect must have a cause—this is generally known as the *Cause and Effect Theory.* Clearly (?), when one arrives in a retrograde fashion to the first effect, whatever that may be, it must be preceded by a cause. Since God is Creator, He must be effect uncaused by a previous cause—and therefore, *uncaused cause.*

This is *theology.* Let us return to rational man. Rational man has difficulty in accepting a god concept and supernaturalism. Consequently, many philosophical and scientific thinkers decline to accept the uncaused cause concept. The study of astronomy (an outgrowth of astrology) has given much impetus to the considerations of cosmology, and the currently accepted concept of the big bang theory is most popular. This, together with the observation of the "expanding universe," has captured the imagination of most astronomers. But let us examine the idea of the "big bang." Inherent in this idea is the premise that "in the beginning" there was a huge mass of matter which exploded, and the effects of this monstrous explosion are still evident (see Fig. 1). *But* where did this huge mass of primordial matter originate? According to theory, it was *uncaused cause.* So we can see, even by avoiding theology, we have come back to the same idea. And it may be that both approaches are *equally irrational.* Is there an uncaused cause?

Now if anybody is so inclined as to propose a set of postulates, such an individual can build a whole system on these postulates. The postulates that lead to *uncaused cause* (what we call theology) are certainly permissible. The fact that such a system includes "miracles" is true only to the observer who is inclined not to accept an action which appears to him to be contrary to the laws of nature. The latter is the definition of a miracle to the rationalist who has not adopted the theologic postulates. The theologian has no problem in accepting as natural what the rationalist considers "contrary to the laws of nature" because, to the theologian, what the rationalist would call a miracle *is* natural and *not* contrary to the laws of nature. The theologian merely admits, quite readily, that he does not understand the laws of nature. As has been shown, the big bang theory really is a variation of the *first cause* concept. Therefore, it too might be considered to be in the realm of theology.

Granted, either of the above concepts could be correct. Both, however, suffer from the same tactical "weakness" in that neither idea was set up as a postulate from which its respective system was built. While it is true that we frequently find it necessary to work *back* from an observation in order to locate its origin, having worked back to such an origin, we ought to be able to set that origin up as a postulate and to build our system. The resulting system may have holes in it by virtue of incomplete knowledge, but it *must be consistent* to qualify as a good system. Experience tells the objective observer

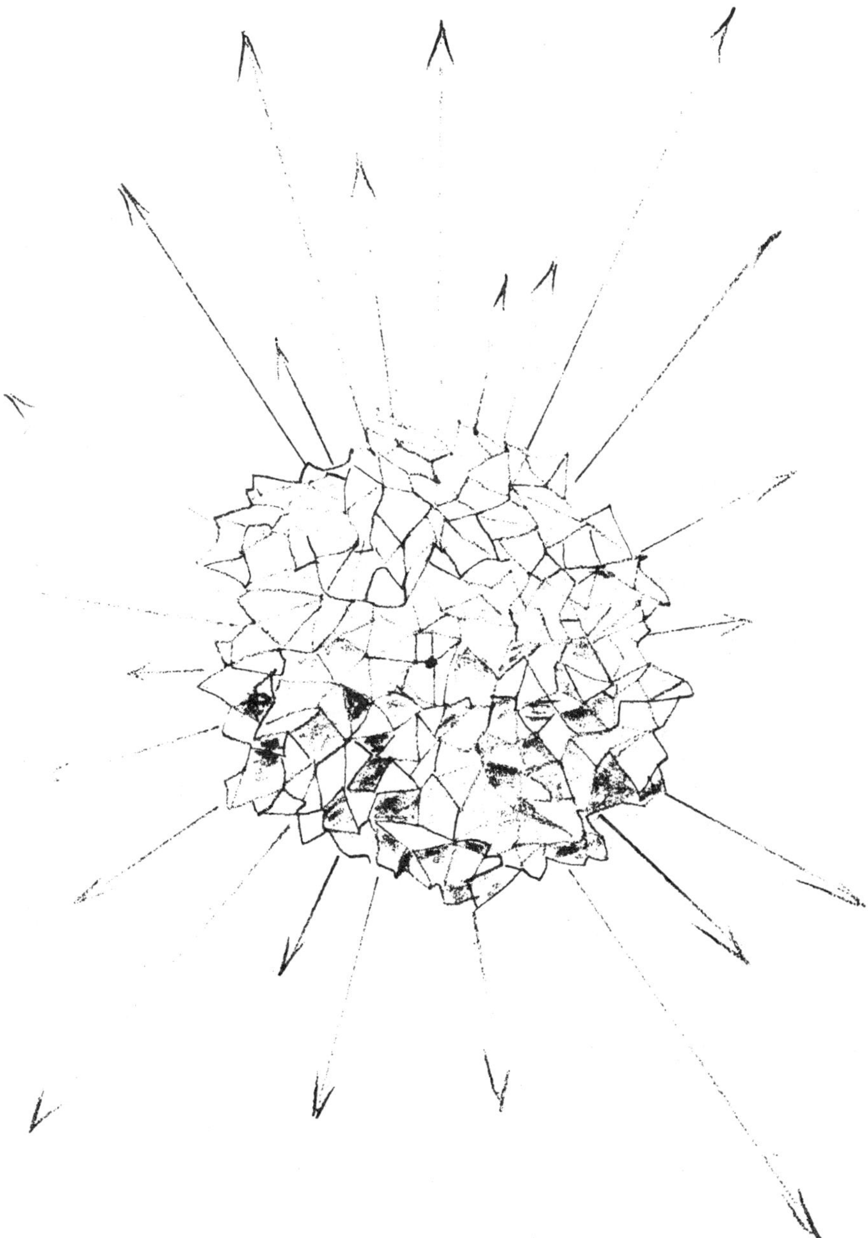

Figure 1. The big bang

that neither version of first cause is consistent. It is for this very reason that the original philosophers and scientists dared to search for answers which were contrary to the theologic norm. This was the beginning of the scientific method.

Clearly, there would be no need for the present exposition if it were not believed that the current systems are inadequate, and if it were not believed that there is a better one to offer. The current presentation, too, results from working *back* from somewhere in the middle. However, it is believed that we can have better consistency with the following.

Though experience can be misleading, we shall propose the following postulates from Man's combined experience:

1. There was a beginning.
2. Matter is that which has mass.
3. Mass is that which, on motion, has momentum and force.
4. Energy is the business of matter; it is what matter does.
5. Energy without matter does not exist.
6. All action must be direct; it is the result of matter in motion.
7. There is no such thing as "action at a distance."

In the beginning, there was nothing. What is nothing? It is the absence of anything, therefore, no thing. Let us contemplate a vacuum–to our notion, the absence of anything–at least the absence of anything with which we are familiar. Let us assume that all of the universe is a vacuum and that, in a manner which we cannot fathom, but surely not supernaturally, what we think of as nothing, splits apart (call it fission) and the result is two particles of matter. One of these particles, which we shall call an *emag* (for electromagnetic particle), will have unit mass. The other particle, which we shall call *other emag,* will have unit *negative* mass. In the process of this fission, the *small bang,* if you will, both particles are imparted with motion in opposite directions (see Fig. 2). The sum of everything is *zero*–nothing! We have deliberately chosen the term *other emag,* or *other matter,* rather than *antimatter,* at this juncture, because *antimatter* has other connotations at this time.

In a haphazard manner, therefore, we have limitless numbers of small bangs throughout the limitless and infinite universe, a process which still goes on. In other words, matter is constantly being created, and the law of conservation of mass (or energy) is not valid in these dimensions. There are almost infinite variations and permutations of aggregates of these emags and *other emags,* and as a result, what we consider as electrons and protons and other subatomic particles are merely the few that we have been able to recognize and demonstrate. These particular aggregates, evidently, happen to be the ones which have greater stability and have yielded to our search. The above process of obtaining something from nothing might take almost limitless time to achieve, of course, but in the overall picture, time is no

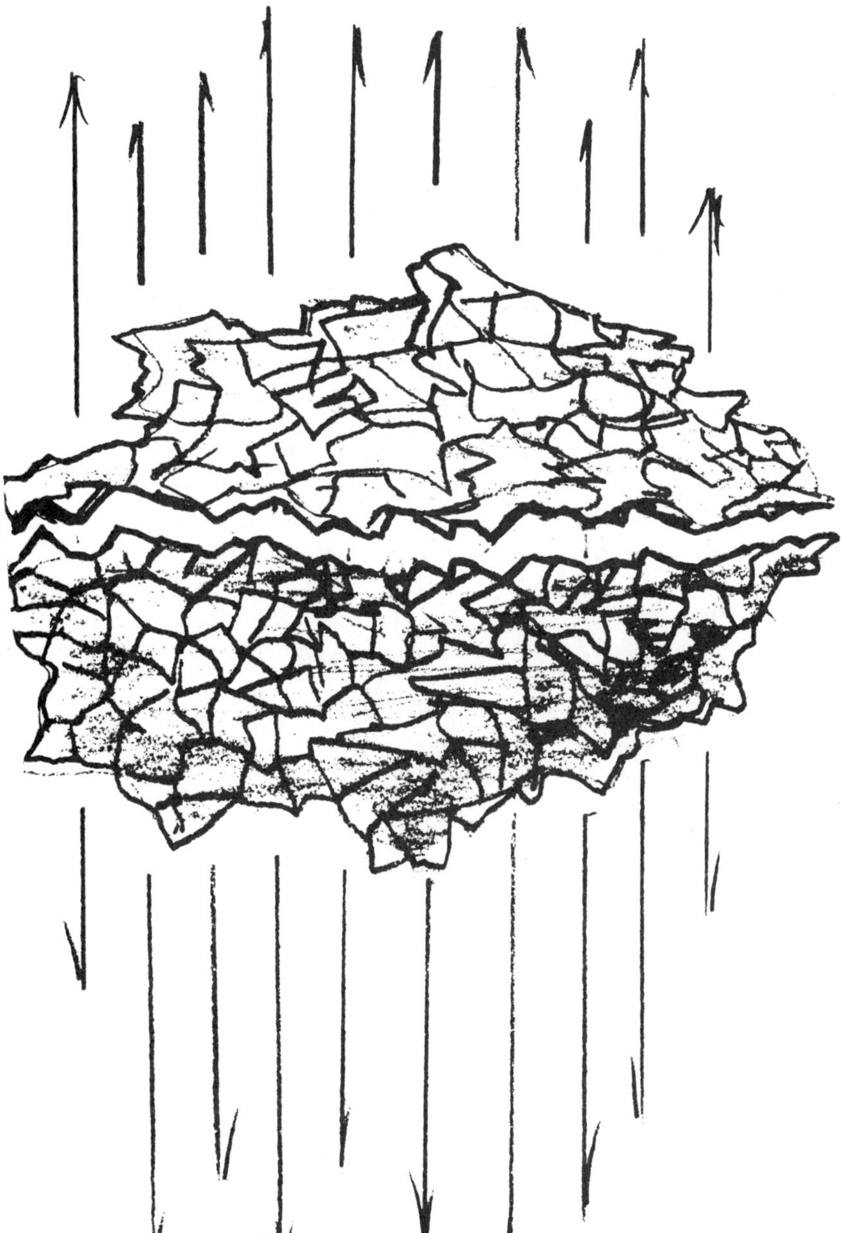

Figure 2. The small bang

restricting factor. Mathematically, and actually, there is no contraindication to negative mass.

It might be argued that, according to the above theory, *nothing* cannot be nothing, since it is composed of two somethings. Where did these two somethings originate? This is true only in the same sense that zero is something. At any rate, we cannot assume that there was always something, since it would have to come from somewhere. We have no difficulty in postulating that originally there was *nothing* and then something happened, in an unknown manner—perhaps as described above. It is also possible, of course, that our "rational concept of cause and effect" is erroneous. However, by accepting the possibility of the small bang theory, we are not yet obliged to discard the theory of cause and effect, which thus far has served us well.

Assuming that the above hypothesis *is* tenable, where does it lead us? Is it compatible with other knowledge of physics? Does it have any advantages over other cosmological theories? We propose to show that the answer to these questions is yes. As indicated above, we now have innumerable emags and *other emags* throughout the universe, after an indeterminable period of time. Haphazardly, there occur collisions between these emags and *other emags,* resulting in the gradual aggregation of larger and larger masses.[1] If an emag of unit positive mass collides with another emag of unit positive mass, and the two become travelling companions, the aggregate is a particle of two units of positive mass. If an emag of unit positive mass collides with an *other emag* of unit negative mass, and the two become travelling companions, the aggregate has zero mass, which is where it all began. If a mass of emags of 10 units of positive mass collides with an *other emag* of unit negative mass, the result will be a particle of 9 units of positive mass, and so on. It is apparent, too, that when one aggregate of emags collides with another aggregate of emags, and they become travelling companions, the conglomerate mass will move off in a direction and with a velocity which is the resultant of the two aggregates. In this way, there is possible almost any number of aggregates and any number of velocities. Is there a limiting size? Probably, asymptotically. Is there a limiting velocity? Probably, asymptotically. Is this limiting velocity the velocity of light? There is no reason to think so! It would appear that electrons and protons are the more stable aggregates of emags, though there should be, and probably are, innumerable aggregates of intermediate sizes and also those greater than protons. What is it that

1. Though, initially, it was tempting to permit (in the theory) collisions only between emags and emags and only between *other emags* and *other emags*, leading to two universes, one of matter and one of *other matter*, this alternative was abandoned for reasons which will become clear. It was equally tempting to consider matter and *other matter* to be mutually repulsive, accounting for the expanding universe. It was believed, too, that there would have to be a positive, strong force to affect a collision between an emag and an *other emag*, since this would be contrary to the natural state causing the two emags to repel each other. The absence of such a strong force, except by coincidence, would contribute to the already expanding universe.

might make the other aggregates less stable? Presumably, the greater incidence of fractures resulting from collisions. This analysis suggests that fracturing is the cause of radioactive degradation, rather than these phenomena being spontaneous processes.

While postulates are arbitrarily accepted "rules of the game," for the "game to make sense," the "rules" or postulates ought to be agreeable and acceptable to the participants. There ought to be no great controversy about accepting postulate #1. Postulate #2 is a reasonable and acceptable definition of matter. Postulate #3 defines the laws of motion. Postulate #4 may be controversial. However, since we are, at this point, "inventing the game" and "making the rules," so be it! Postulate #5 is an extension of postulate #4 and is crucial to the "game." We are inclined, at times, to speak of units of energy as though they were self-sufficient. However, disembodied energy cannot exist. Energy is what matter does, in motion. Postulates #6 and #7 might be considered irrational, since we see "action at a distance," regularly. We see electric "fields"; see magnetic "fields"; and see gravitational "fields." Besides, all calculations based on the field theories work![2]

Consider the alternative: The calculations of "field forces" are mathematically correct, since they *do* work. However, they show what and how much an effect will be and not how it happens. The "how," we suggest, is the effect of the surrounding "sea of emags," which are too tiny to be demonstrated individually. Postulates #6 and #7 are not directed at the question "How much?" They are directed at the question "How?" Field equations explain the former. Emags explain the latter. Let us now put to rest the obsolete *vis viva.*

2. Before postulate #7 was entertained, that is, "no action at a distance," since it was believed that charged bodies have supernatural powers of forces acting without an intervening medium, it was considered essential that emags and *other emags* have charges. There were several options. One option was that an emag or an *other emag* have a single charge, such as an electron or a magneton. This created problems which could not be resolved satisfactorily. Another option was that each face of an emag or *other emag* have a separate charge. Both options were discarded in favor of postulate #7.

2. What Is an Emag?

To this point, we have concluded that the universe comprises only emags, *other emags,* and aggregates and combinations of these. Can we propose a description of an emag? Yes. To do so, we shall postulate the *Economy of Nature,* that is, a variation on Occam's razor. Nature evolves in such a manner as to result in the accomplishment of a task by the simplest route. We have postulated the emag as a particle of matter. It has volume and mass. Furthermore, since it is the ultimate building block of the universe, it is incompressible. We shall postulate, too, that this emag must be a symmetrical solid. What might it look like? Following the lead of the *Economy of Nature,* let us try various possibilities. Can we make a symmetrical solid from a single plane? No. Can we make a symmetrical solid from a single curved surface? Yes, a sphere. Can we build symmetrically with spheres? No—there are always vacant spaces between them. Can we build a symmetrical solid from two curved surfaces? Such a solid would not be universally symmetrical, and building with them would still result in empty spaces. How about two planes? No solid. Three curved surfaces? The same objections apply as with two curved surfaces. How about three planes? Again, no solid. Four curved surfaces? This is interesting. We can have four convex, spherical triangles, but we cannot build without empty spaces. The same is true for four concave, spherical triangles. If we use a combination of convex and concave spherical triangles, we might be able to build without empty spaces, it would appear (actually, we can't: try it). At any rate, such a combination would not be universally symmetrical. Well, then, what about four planes? A regular tetrahedron has all of the qualifications we have proposed. Each surface is symmetrical, and just like any other surface of the solid. Also, we can build anything with these tetrahedrons, without vacant spaces. Let us, then, propose that the emag is a regular tetrahedron (see Fig. 3).

Of course, there is no *a priori* need for a symmetrical solid. If a ragged chunk of matter should happen to break off from "nothing," as an emag of unit mass, the remaining ragged chunk would be a unit of negative mass. It is perfectly conceivable that various, different, haphazard chunks of matter

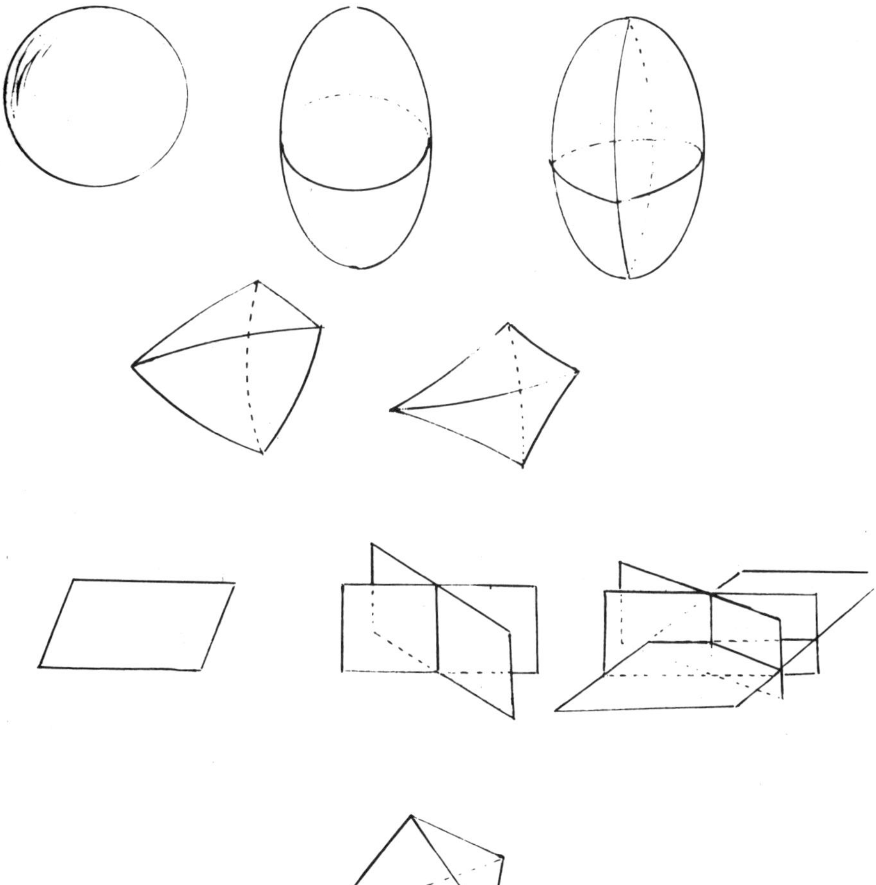

Figure 3. Possible emag shapes

should break off from "nothing" and there be no uniformity of emags and *other emags,* let alone symmetry. Furthermore, there is no real need for emags and *other emags* to build aggregates *without* spaces between them. While any of the above alternatives are permissible, and could actually be the reality, since we are inventing postulates, it is convenient to demand symmetry, uniformity, and aggregation without spaces, as the following will show (see Fig. 3).

Where can we go from here? Let's build electrons and protons. Once again, we shall make some assumptions. We shall assume that both electrons and protons are spherical bodies, without crevices, and, therefore, have the same specific gravity. These assumptions may not be true, but are made merely for convenience, in order that we may seek to determine the dimensions of the emag. With the above assumptions and the knowledge that the mass of the proton is approximately 1836 times the mass of the electron, the following development holds.

Assume:

1. Emag is regular tetrahedron.
2. Electron is solid composed of emags, without crevices.
3. Proton is solid composed of emags, without crevices.
4. Therefore, electron and proton have same specific gravity.
5. Electron and proton are spherical.

1. $$\frac{\text{Area sphere}}{\text{Volume sphere}} = \frac{4\pi r^2}{(4/3)(\pi r^3)} = \frac{3}{r}$$

2. In order to "neutralize"[1] a proton with an electron, half the surface of the proton needs to be covered by the emags in an electron. One-half the surface of the proton would need to be covered by enough emags so that one face of the electron's emags would cover one face of the proton's surface emags. Since all the emags in the electron need to be used, the surface area of the proton, divided by the area of one face of the emag, must equal twice the number of emags in the electron. Or,

3. $$\frac{\text{Area proton}}{\text{Area of emag face}} = \text{numerically, } 2 \times \text{number of emags in electron}$$

4. $$\text{Number of emags in electron} = \frac{\text{Volume electron}}{\text{Volume tetrahedron}} = \frac{VE}{0.11785\,L^3}$$

5. Since area tetrahedron face is $0.43301\,L^2$,

1. This analysis is based on "charged" protons and electrons. Of course, since we have abandoned charged bodies (see page 4) it is not correct. It is included in the essay as a rough indication of the order of magnitude of the emag.

6. $\frac{\text{Area proton}}{\text{Area of emag face}} = \frac{2VE}{0.11785\ L^3}$

7. $\frac{\text{Area proton}}{\text{Volume electron}} = \frac{2 \times 0.43301}{0.11785} = 7.3485$

8. $\frac{AP}{VE} = \frac{4\pi r_p{}^2}{\frac{4\pi r_e{}^3}{3}} = 7.3485$

9. Since mass of proton is known to be 1836 times the mass of electron, and both are spheres of the same specific gravity, the radius of the proton is $\sqrt[3]{1836} = 12.245$ times the radius of the electron (see Fig. 4).

10. From 8 : $\frac{4\pi r_p{}^2}{\frac{4\pi r_e{}^3}{3}} = \frac{(12.245 r_e)^2}{\frac{r_e{}^3}{3}} = 7.3485$

11. $12.245^2 \times 3 = 7.3485 r_e$

12. $r_e = 61.212 = 6.1212 \times 10^1\ L$, where L is edge of emag.

13. $r_p = 7.4954 \times 10^2\ L$

14. $V_p = \frac{4\pi(7.4954 \times 10^2)^3}{3} = 1.7639 \times 10^9 L^3$

15. $A_p = 4\pi(7.4954 \times 10^2)^2 = 7.0599 \times 10^6 L^2$

16. $V_e = \frac{4\pi(6.1212 \times 10^1)^3}{3} = 9.6072 \times 10^5 L^3$

17. $A_e = 4\pi(6.1212 \times 10^1)^2 = 4.7085 \times 10^4 L^2$

18. Number of emags in E (4) $= \frac{9.6072 \times 10^5}{0.11785} = 8.1520 \times 10^6$

19. Number of emags in P is $1836 \times E = 1.4967 \times 10^{10}$

It must be reiterated and emphasized, at this point, that the above analysis is not correct, since the five assumptions are not necessarily valid. However, as an exercise in orders of magnitude, it may have merit. Additionally, since the mass of the electron is 9.1091×10^{-31} Kgm, the mass of the emag, according to the above analysis, is of the order of

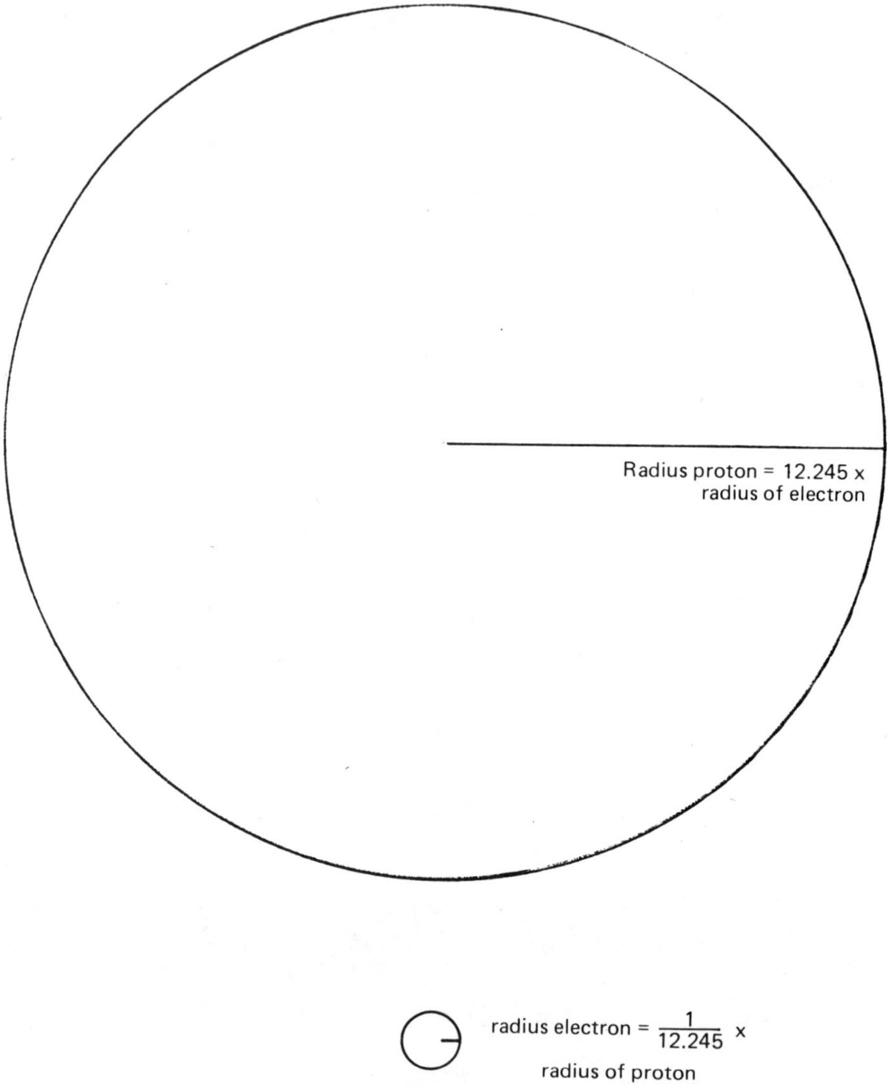

Figure 4. Radius of proton is 12.245 times radius of electron

$$\frac{9.1091 \times 10^{-31}}{8.1520 \times 10^{6}} = 1.1174 \times 10^{-37} \text{ Kgm.}$$

Also, if the radius of the electron is approximately 2.8178×10^{-15}m[2] then, from (12) *L,* the edge of the emag, is

$$\frac{2.8178 \times 10^{-15}\text{m}}{6.1212 \times 10} = 4.6033 \times 10^{-17}\text{m}$$

$$= 4.6033 \times 10^{-2} \text{ Fermi}$$

See Fig. 5.

Let us return to our five assumptions, but not in the order presented. Are the electron and proton spherical? It is a convenience to think so, but this is surely open to question. Do the electron and proton have the same specific gravity? This is not hard to accept, but it need not be true. Are the electron and proton solids, composed of emags without crevices? We are obliged to accept the composition of protons and electrons as emags because this is a basic postulate of the emag hypothesis. However, we are not obliged to accept the idea of the absence of crevices between the emags. Furthermore, we are not even obliged to assume that the composition of the electron and proton is restricted to emags. There is every reason to suspect that all subatomic particles are aggregates of emags and *other emags,* in a haphazard fashion. A combination of emags and *other emags* would merely leave massless holes in the larger particles. It is not unreasonable to assume, by the laws of chance and probability, that the average percentage of massless holes is approximately the same in both the electron and the proton, thus maintaining a constant specific gravity. And now, the initial and basic assumption: is the emag a regular tetrahedron? By the principle of the *Economy of Nature,* such a supposition seems reasonable. However, how would such colliding emags aggregate? It is tempting to ascribe electrical and/or magnetic charges to these tetrahedrons, in order to hold them together. It would be necessary to ascribe "holding power" and neither attractive nor repulsive power to these charges, in order not to run afoul of "action at a distance," which we have decided to abandon. In other words, the charges would act like glue, on contact. This, however, appears too contrived–particularly since we would be hard put to explain the action of like charges, in contact with one another.

Electric and/or magnetic charges on the emags do not seem to be of value. Well, then, in the absence of some "holding power," how *do* colliding emags stay together and aggregate to larger masses? Despite the principle of the *Economy of Nature,* sometimes the simplest route does not work! Under such circumstances, we are obliged to try the second simplest explanation.

2. *Handbook of Chemistry & Physics,* 49th Edition, 1968-1969, page F-167.

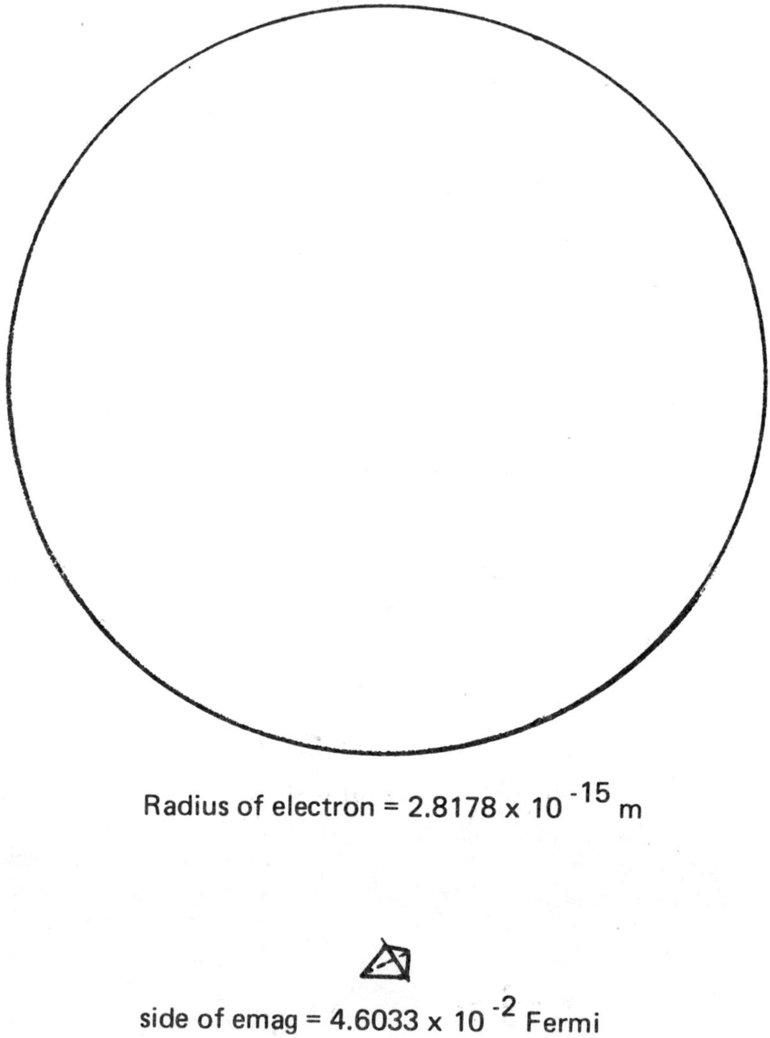

Figure 5. Relative sizes of emag and electron

We must bear in mind that colliding bodies do not stay together, but recoil from one another. This kind of collision will not permit emags to build into larger aggregates. Suppose, however, that our original particle of "nothing" (no mass) has the form of a regular tetrahedron, and that the small bang results in the separation of the core of this particle in such a manner that the core, in the form of a smaller regular tetrahedron, has an equal and opposite mass to the shell of the original particle. The sum of the two emags is still zero, but now we have two different kinds of emags, i.e., the core and the shell, either of which could be the emag, with positive mass, and the other being the *other emag* of negative mass. Now we have possibilities. On collision, the apex of one shell (emag or *other emag*) can impact itself into the empty space of another shell (emag or *other emag*) and build chains.[3] These chains, as they increase in length, can now twist and spiral and coil, and such a complex structure can be stable. Nor need the electron or the proton be restricted to one such chain. There might be the complex intertwining of several or multiple chains. A common example of this, in daily life, is the aggregation of strands into balls of tumbleweed. Growth may occur at either end of the chain, but the "female" end could be filled with a smaller tetrahedron. This concept provides us with another "dividend": as the shells grow into chains, there remains a large excess of smaller tetrahedrons, both emags and *other emags*. These constitute the "sea of emags" to which is attributed the "field" forces, and, in the emag hypothesis, all direct action (see Figs. 6, 7, and 8).

3. We are not obliged to have smooth surfaces for our emags or other emags. Jagged surfaces would assist in "togetherness."

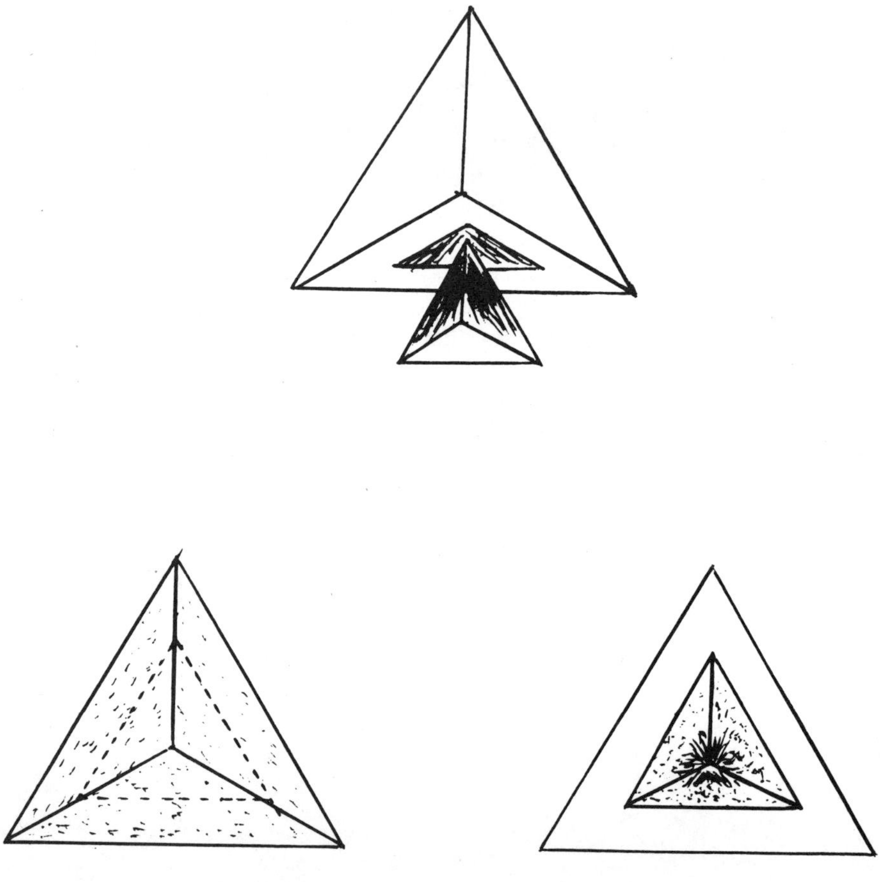

Figure 6. The emag and *other emag* as core and shell

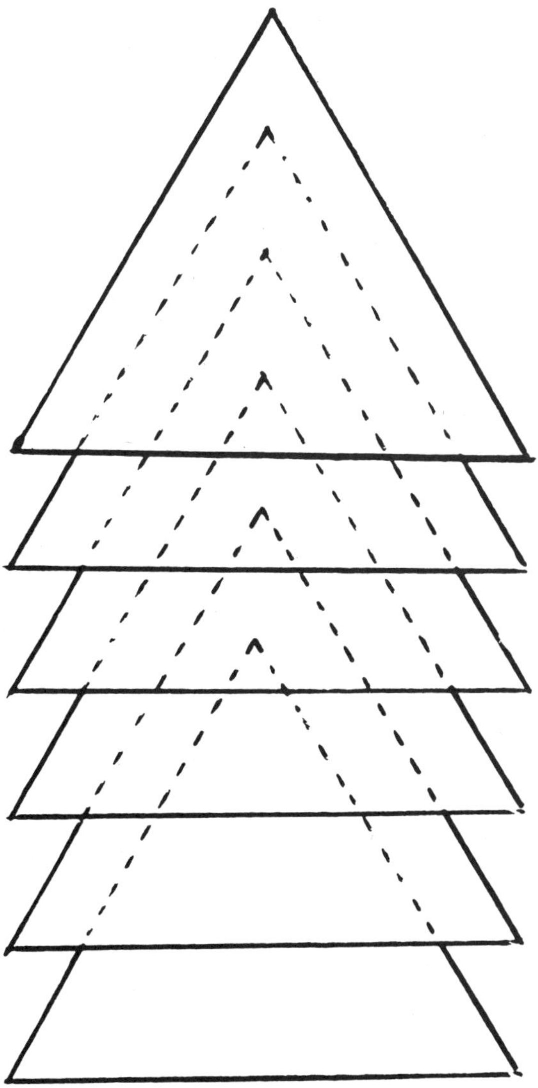

Figure 7. Emag and/or *other emag* shells forming strands

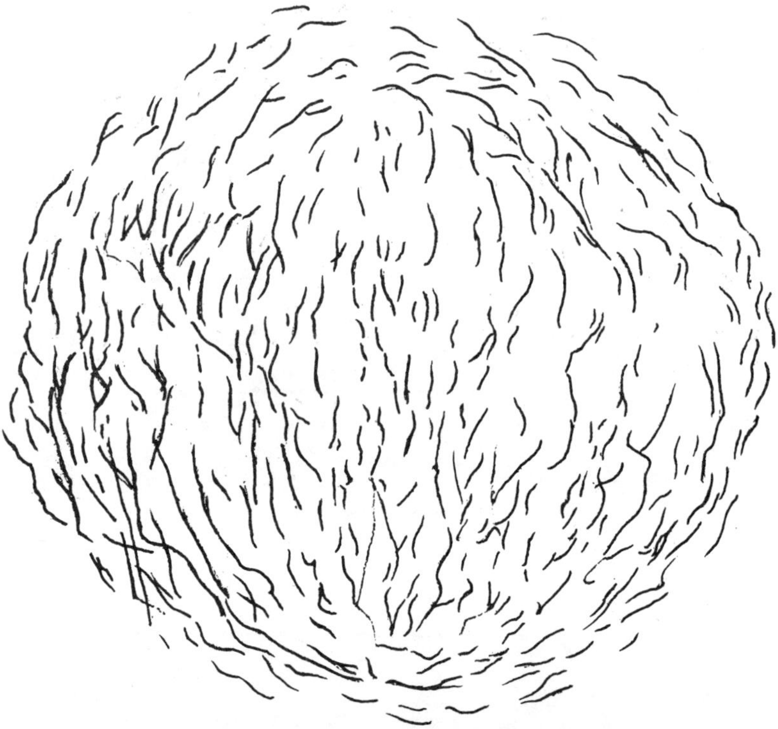

Figure 8. The "tumbleweed" concept of protons and electrons

3. Extension of the Emag Hypothesis

In quantum theory, the atom is viewed as a central nucleus surrounded by "electron clouds." In the Bohr atom, the electrons are viewed as surrounding the central nucleus in definite orbits. Clearly, in the case of hydrogen, according to the Bohr theory, the electron orbits in a plane, which may precess. In the case of the helium atom, there is the presumption that the two electrons orbit in a plane (which may precess). As a matter of fact, by the Pauli exclusion principle, the two electrons are essentially in the same orbit. Whether by the Bohr theory or the quantum theory, as the number of orbiting atomic electrons increases, the presumption is that the electrons are not in the same plane, by orbit or cloud, but, rather, distribute themselves in three-dimensional space, around the nucleus. This would surely be anticipated as a rational expectation, and additionally, the electrons "repel" one another, and so *must* distribute themselves in three dimensions.

How does the above analysis fare with the emag hypothesis? Since the emag hypothesis does not provide for "action at a distance," neither repulsion nor attraction, there is surely no need for three-dimensional distribution. Furthermore, if all orbital motion is due to the *direct action* of the push of the emag sea, we may presume that in *any* planetary system, small or large, the emag sea has a major direction, or current, associated with lesser ones. Each planetary system, if stable, and we presume it is in the context of its own time scale, *must* have such an arrangement. Now, if we are dealing with a major emag current, should we not expect the orbiting electrons to be, essentially, in the same plane, but at different distances from the nucleus? Of course, we do not know whether or not this is so, but let us seek some parallels. The most natural and obvious planetary system to observe would be our solar system. And, lo and behold, with the exception of Pluto, the other planets do lie, essentially, in the same plane–the ecliptic (see. Fig. 9)!

But why is Pluto an exception? Note the distance of Pluto from the sun. Evidently, Pluto is not in the same "stable" planetary system as our other planets, around the sun.

Can we find another parallel? But of course. Any spiral galaxy, including our own Milky Way, is such a system. But how can we explain nonspiral

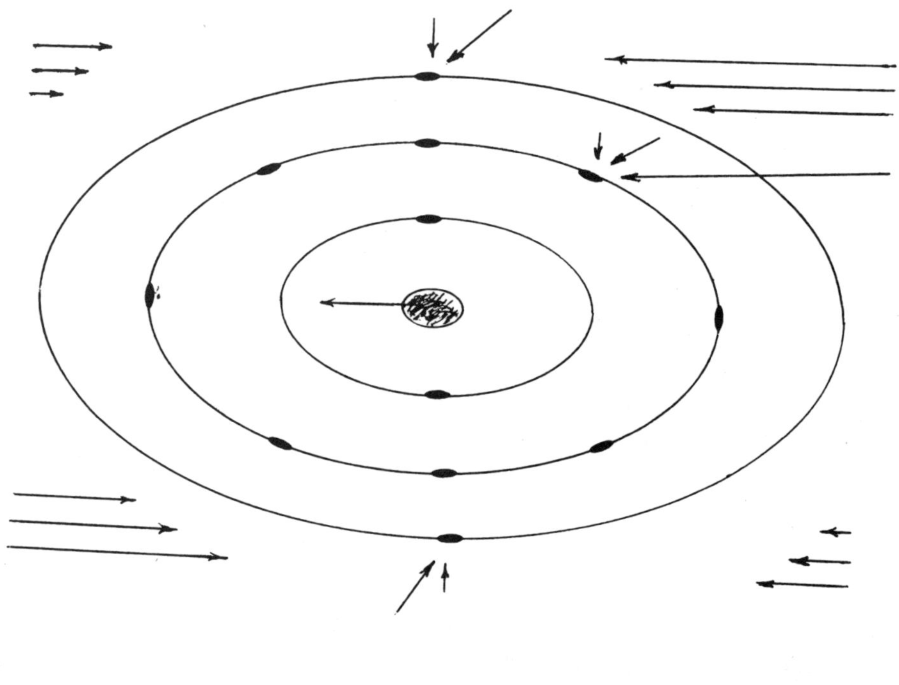

Figure 9. Electron Paths

Elliptical orbital paths of electrons are due to the resultant forces of all the emag currents in any particular vicinity. As the arrow at the nucleus shows, the major resultant force is to the left. Different resultant forces at different locations give orbital motion.

or globular galaxies? For one thing, we have no knowledge of the "depth" of such galaxies. They may be several "overlying" galaxies, with huge distances of separation. Furthermore, they are at such great distances from us (of the order of comparison of the solar system to the atomic system) that we may have no concept as to the emag current there. As a last resort, we may even have to concede, if the above explanations should be inadequate, that we simply do not, as yet, have all the answers.

As an aside, it is interesting to note that our moon must also lie, generally, in our solar ecliptic, since monthly we have a "new moon", indicating that the moon and Earth and sun are, essentially, in line, barring precession. Of course, with a solar eclipse affecting some part of the Earth's surface, we are acutely aware of the direct line.

Let us turn our attention to the so-called "expanding universe." When (?) the primordial mass exploded, the fragments, large and small, were thrown out, radially, in all directions. Let us examine this notion. If an observer were at the center of the universe, all would be clear. If this observer were to look in any direction, he would see masses receding, radially, without crossover. We on Earth, of course, are in relative motion, rotating around our own axis, revolving around the sun, and moving with the sun through space, apparently, in an orbital manner in the Milky Way, which, in turn, is also moving through space in some fashion. So we are contending with at least four different motions. How should the universe look to us? It is safe to say, categorically, when we consider man's life cycle in relation to the time scale of the universe, that we shall never know.

But astronomical observations and Hubble's law indicate an expanding universe. In view of our previous statement, what *are* we seeing? We are seeing what we *think* we are seeing, secondary to our *original bias* about what the expanding universe should show us! But, you might say, weren't the astronomical observations the bases for the proposal of the big bang theory and the expanding universe? Ideas are cyclical. The thought occurs; the observations are made; if thoughts and observations seem to mesh well, further thoughts and further observations are made, and attitudes and biases develop automatically. For example, the heliocentric system of Copernicus was not his brain-child. The idea was ancient and had been discarded, essentially. Copernicus's studies of history, and his observations, led him back to the heliocentric bias. Similarly, with the big bang theory. The idea had been suggested, making the expanding universe axiomatically essential, and when other astronomical observations, in the presence of such a bias, were felt to be compatible with the big bang, the cycle was strengthened.

However, let us return to the "at least four different" motions of the Earth. If the big bang theory were a fact, as we have indicated previously, we could not know what the result should look like from the multiple motions of the Earth. So, what we are *seeing* and interpreting as the result of the expanding universe may have another and better explanation.

We must return to the basic small bang hypothesis. Somewhere, sometime, in multiple "places" in the original nothingness, small bangs occurred, resulting in emags and *other emags.* Since the original nothingness had no unique points, every point was equal to any other point. Just as we cannot predict, in advance, which radioactive atom will disintegrate, so there was no preference as to where a small bang might occur. Just as we can predict the rate of radioactive disintegration, there was *some* unknown rate of small bangs occurring throughout the original nothingness. Such a process would occur anywhere and at any time, according to some probability time scale. Since there was no preferential time nor place, space, as we speak of it, automatically became infinite! There can be no end to space–not only boundless, but infinite.

The agglomeration of emags which has resulted in the universe is obvious. And now, we are confronted with another obvious, almost incongruous conclusion: *Any place is the center of the universe,* since any place is infinitely far from the "edge" of the universe. Can we, then, assume that we are at the center of the universe and explore the heavens from this vantage point? Yes and no. If we are to assume that we are central, this implies that all motion is *dependent* on our central position. This is clearly not the case. Also, since any other place can be considered as the center of the universe, all motion would also have to be *dependent* on *that* central position. This is, also, clearly not the case.

Well, then, what can we do to evaluate motion? We must accept the obvious conclusion that all motion is *not* dependent on *any* central position—not even the presumed central position of the big bang. But what is it we are seeing which we are attributing to the expanding universe? We are seeing *some kind* of motion, with some kind of recession (the red shift–if we accept the Doppler effect), but *not radially,* necessarily, and only radially, at times, by chance. For example, if we were to observe the red shift of any distant star and attribute it to radial motion, it might very well be due to curved motion, the radius of curvature being unknown and huge.

Apparently, any body involved in rectilinear motion does so only by chance, and no observer could know it, since he himself is in relative motion. Rather, almost all motion is curvilinear motion. What is the impact of this on Newton's laws of motion? Quite little. As a matter of fact, it is only when the curvilinear motion clearly has a relatively small radius of curvature that we need concern ourselves with nonrectilinear motion. What is the impact of the above on the rate of "expansion" of the universe? From time to time, there have been reports of the discovery of a celestial body which is travelling faster than the velocity of light. Such reports have never been confirmed. Nor is it likely that they ever will. Since the determination of the velocity of a celestial body is dependent on the measurement of the red shift, any time such a body is receding from us with a velocity greater than that of light, we will never know it, since the light never reaches us, unless, of course, we should discover another means of measuring velocities greater than the velocity of light. If we should ever discover a body whose velocity of recession is equal to the velocity of light, there will be *no Doppler shift.* This statement, of course, negates Einstein's second postulate. Can there be any motion faster than the velocity of light? There is no reason to assume that there cannot be, and there is no hard·evidence that there is not. The only thing we can say is, as before, that if there were such a motion of a body, we would not see it.

4. The Michelson-Morley Experiment

We have taken the position that the *ether* is real and that it is the *sea of emags*. This, of course, requires "squaring" with the Michelson-Morley experiment. Now the Michelson-Morley experiment "proved" the absence of the ether and was central to the development of Einstein's second postulate. The experiment was set up, initially, to study the ether; but the absence of fringe shifts suggested that there was no inherent ether velocity–i.e., no ether wind, ergo, no ether. On re-evaluation of this experiment, in the light of what we have been saying, we find the experiment to be evidence *for* the emag hypothesis. Consider our comments about the rotation of the hydrogen electron around its proton nucleus. It is due to the push of the emag sea. We have postulated the absence of *action at a distance*. Consequently, it is not only the hydrogen electron which is being pushed around its nucleus by the emag sea, but *all of the orbiting heavenly bodies*. In other words, the Earth, orbiting the sun, is kept in motion by the push of the emag sea, and it is "riding" in the sea. There is no gross relative motion of the emag sea *past* the Earth, and so the interferometer *should show no fringe shift*. This is comparable to the Bohr stationary state.

5. *Relativity and the Second Postulate*

As mentioned earlier, the "absence of the ether" implied that the velocity of light is constant throughout space, and unaltered by the motion of its source. This statement, Einstein's second postulate, is now suspect. Additionally, the Doppler effect indicates that the velocity of light *is* dependent on the velocity of its source. Since a postulate is a statement which is accepted as true whether or not it sounds reasonable, how can we question it? We find it more rational to accept as a postulate a statement that seems reasonable, rather than one that does not. That the velocity of light is constant, ultimate and independent of the velocity of its source, is challenged by the Doppler effects of red and blue shift, and also by other considerations.

Surely, the source of the light is not changing its frequency. The periodicity of the light's emission remains essentially unchanged so long as there is no change in the *energy state* of the orbital electrons. What does happen is that if the source of the emission is approaching us, the time it takes for the light to reach us is shortened by the contribution in velocity by the source. As a result, the same number of impulses reach us in a lesser period of time. This is the same effect as though a greater number of impulses reach us in the same period of time—the blue shift (see Fig. 10). If the source of the emission is receding from us, the fixed number of impulses take longer to reach us, by virtue of the source's motion away from us. The result is a lesser frequency per unit of time—the red shift.[1]

1. An interesting experiment can be performed. Strike a tuning fork and hold the base against the bone behind the ear. When the sound disappears, move the tuning fork to a position beside the ear. The sound returns and is heard again. This demonstrates that air conduction of sound is better than bone conduction. However, though the middle ear is a mechanism that is more sensitive to sound than is the bone of the skull, sound travels *faster* in bone than in air. Now, reverse the order above. Strike the tuning fork and hold it near the ear. The tone is heard. Move the base of the tuning fork to the bone behind the ear and the tone is found to have a slightly higher pitch.

Let us analyze this. The tuning fork has a fixed frequency, yet the pitch, which is a measure of frequency, is higher with bone conduction. Clearly, in any unit of time, the same number of cycles of sound is carried a greater distance in bone than in air. This is the same as a greater number of cycles in a given time. This is the Doppler effect. In

(continued on page 27)

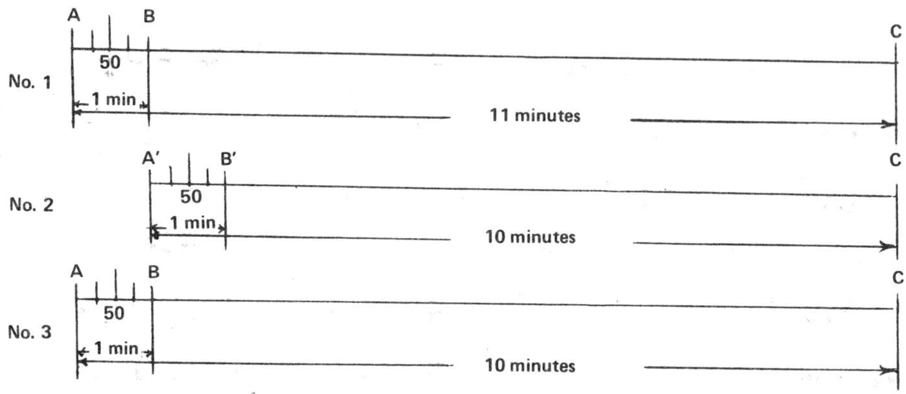

Figure 10. The velocity of light is dependent upon the velocity of its source

1. Oscillator sends 100 pulses per minute, starting at A. It takes one minute for 100 pulses to move from A to B. It takes ten minutes for first pulse to reach C from A. Therefore, it takes eleven minutes for 100th pulse to reach C from A.

2. Same oscillator as #1. Start is A'. It takes one minute for 100 pulses to move from A' to B'. It takes nine minutes for first pulse to reach C from A'. Therefore, it takes ten minutes for 100th pulse to reach C from A'.

3. Same oscillator, but now oscillator is moving from A to B in one minute. During that minute, 100 pulses are sent. Thc first pulse arrives at C from A in ten minutes. By the time A reaches B, the 100th pulse has been sent, and it will take the 100th pulse nine minutes to reach C from B. Therefore, the 100th pulse will have reached C nine minutes after the first pulse. The effective frequency of the oscillator will then be 10/9 "times" 100, or 111 pulses per minute. The effective velocity of the pulses is ten minutes from A to C, as opposed to 11 minutes in #1, where the oscillator is not moving. Velocity is dependent on motion of the source.

How do we measure the velocity of anything? We determine the distance between two points and then determine the time it takes to traverse that distance:

$$v = d/t$$

How do we measure the velocity of light? The Michelson method does exactly that. So we know the velocity of light between two points *on Earth,* the frame of reference of the experiment.

According to Einstein's second postulate, if two objects are in motion in space, in opposite directions, and if an observer in one of these objects fires a flare at the moment that the two objects are adjacent, that flare being fired from the momentary common origin of the two frames of reference of those two objects, an observer in each of those objects will "see" himself as the center of the expanding flare of light, even as the moving objects continue to separate. Now it may be well and good to say this, *but* this demands that the center of the flash be at two different points at the same time, since the objects have moved apart. Furthermore, if, when the two objects are adjacent, each observer fires a flare, each of the flashes will continue to travel with that observer, even when the objects have separated, and each observer will "see" himself at the center of his own flare. Surely, the two circumstances are not identical. While in each instance the observer on each object may *believe* he is remaining at the center of the flare, in the first example, one is, unknowingly, in error. In the first instance, not only is one observer's *individual impression* in error, how could he prove his contention? A third, neutral, observer, looking in from the "outside," clearly does not see two expanding flashes of light in the first instance, and clearly does see two flashes of light in the second instance.

Let us measure the velocity of light between the Earth and the moon, assuming they remain a constant distance apart. In order to do so, we must know that distance, which we do not know, with accuracy. Our knowledge of that distance is based on the premise that a laser beam, reflected from the moon back to Earth, is travelling at the *recognized* velocity of light. In other words, we cannot measure, with accuracy, the velocity of light anywhere, except on the Earth.

But let us assume that the velocity of light *is* constant, though we shall demonstrate that it is not. We can, by no means, claim that if a star is sending

other words, if an oscillator of fixed frequency is moving towards us, that freqency covers a distance greater than its own velocity would accommodate, per unit time, by an amount dependent upon the velocity of the motion of the oscillator towards us. If the oscillator is the atoms of a star, and that star is rotating so that one side of its circumference is approaching us, the effect is the blue shift, *dependent upon the velocity of the source.* The apparent frequency has increased because the source is travelling faster than it would be if the rotation of the star were at right angles.

light to Earth, then the time of transit of that light is independent of that star's relative flight from Earth. If we could measure the velocity of such a beam of light from some point in its flight to some other point in its flight, by virtue of a marker, we might find the velocity of that light to be constant. But we have no such reference points. The only other approach is to "shut off" the light from the star, at its source, and measure the time it takes for the light to reach the Earth, after we "turn it on" again. Obviously, this is impossible. And even if it were possible, we would have to know, with accuracy, the distance of flight from the star to the Earth—another impossibility.

What we are talking about is comparable to the "time of flight" experiment with the electron gun. We must have a starting switch and a marker on the oscilloscope, a sensing switch, and a known distance. If we could do our above-mentioned impossible experiment, we would expect, as the star recedes from us in this "expanding universe," that the light coming from it would take *some* time to recover the distance the star has moved from its starting point, and therefore, the light would take longer to reach the Earth (despite an assumed constant velocity of light) and the *real velocity of light coming from that star would be reduced, giving us the familiar red shift!* It appears, therefore, that Einstein's second postulate is untenable.

A few more examples might be interesting and instructive. These have to do with the concept of *absolute time,* as advocated by Newton, versus *relative time,* as advocated by Einstein. It is here proposed that Newton's time is more acceptable. *Time* is inviolable and it is infinite. *The time* is relative. What is a lifetime for a fly is but a few weeks for man. What is a lifetime for man is insignificant in the greater picture. While we may consider *time* as absolute, *the time* is measured by clocks. These clocks are geared to time as man reckons it, that is, the relationship of the Earth's rotation about its own axis, and the orbit of the Earth around the sun. When Einstein suggests that clocks in motion run slow, it is true only if the observer is not near the clock. Anyone who has used a camera knows that motion can be "stopped" by moving further away from the subject to be photographed. Clearly, *time* has not changed by moving farther from the moving subject, nor has the velocity of the moving subject changed; but the farther we are away from the clock (which stays with the motion), time *seems* to slow down. And so, if a clock on a star is visible from the Earth, *the time* on the clock, as seen by an Earth-bound observer who is watching the star move away, would show the passage of time slower than a similar, synchronized clock at the side of the observer. This effect is simply due to the fact that the velocity of light *is dependent* on the motion of its source.

If an observer races away from a clock with the velocity of light, whatever time the clock shows when the observer starts will remain unchanged, since light from the clock leaves the clock at the same time the observer does, and light from the next moment will not leave the face of the clock *until the next moment* has elapsed. Therefore, the new time on the face of the clock will never catch up to the observer. *The time* stands still, but *time* marches on. The observer may not age with respect to the clock, but he will surely age with respect to *time.*

If the above clock takes off at the same time as the observer does, but it travels at half the velocity of light, in the same direction as the observer, as *time* elapses, the clock is drifting backwards, with respect to the observer. Equivalently, if the observer races away from a stationary clock at half the velocity of light, since light from the face of the clock leaves it with the velocity of light, after the observer has travelled ten minutes, the face of the clock will show the passage of five minutes of time. *The time* has slowed down, but *time* is unchanged.

In the same vein, if we should look, at night, to the western sky and determine that a particular star is racing away from us with a velocity of 0.6 c, and if we then turned to the eastern sky and determined, once again, that another star was racing away from us with 0.6 times the velocity of light, and if the two stars are assumed to be essentially in a straight line with the Earth, star A would clearly be receding from star B with 1.2 times the velocity of light. If we should inform observers on both stars A and B of our observations, and they, therefore, know of each other's existence, would each consider the other to be a *black hole,* since they cannot hope to see each other?

We conclude, consequently, that though *time* is independent of the motion of its source, the velocity of light is source dependent. Does this mean that $E \neq mc^2$? Yes, to the extent that it is only an approximation. And, to this point, it is the best approximation we have, and *it works.*

How does all this affect the Lorentz transformation? If the velocity of all "electromagnetic radiation" is constant, as presently assumed, there is no effect at all. There remains doubt that all such radiation does have a constant velocity (in *vacuo*). If it should not be true, the Lorentz k $(1/(1\text{-}v^2/c^2)^{1/2})$ is not a valid k, and an inaccuracy has been introduced, by Einstein, into his equations. Now it may be, if c is constant (?), even if it is not maximal (which it may not be), that the error introduced is of no great practical moment. Einstein's introduction of special relativity has not altered the utility of Newton's laws in the macroscopic world. And it may be that the inaccuracy we have indicated, if present, does not alter the utility of Einstein's equations in the submicroscopic world. In the world of emags, however, it may be a different story. But the world of emags has no current effect on *how much* things work. Its present importance is in understanding *why* things work.

6. The Corpuscular or the Wave Theory?

The controversy still rages, but in muted tones, as to the corpuscular versus the wave theory of electromagnetic radiation. The current compromise, of course, is that it is both. There is no mistaking the corpuscular aspects of the photoelectric effect, nor of the Compton effect. Is there any mistaking of the wave phenomena of light (electromagnetic radiation)? We observe refraction, diffraction, interference phenomena, and polarization. We can also measure the "wave length" of light. What is more, we can measure the wave length of particles, such as the electron. But what are we measuring? We are measuring the *wave manifestations* of hordes of particles–emags and emag aggragates. Surely, we do not doubt the molecular nature of water when we see water waves. Why should we doubt the particulate nature of light when we see light wave manifestations? As a matter of fact, the Davisson-Germer experiment exactly proves this contention; for a stream of particles (electrons), travelling not in a perfect straight line of one electron's thickness, is reflected from the nickel crystal over a tiny, finite area, and is crisscrossed to give interference phenomena, which are readily demonstrated.

Well, then, what about the diffraction of light? Why does light diffract in passing through a slit? If we consider the dimensions, a slit may be very small to us, but to a stream of emags or "photons," it is a huge tunnel of great length. "Photons" entering one end of the tunnel, in haphazard streams, are reflected back and forth from the sides of the tunnel, and exit, again, in a crisscross manner. The pattern of exit is determined by the angle of entry, the diameter of the tunnel, and the length of the tunnel. The exiting "photons" collide with one another in a fanned-out beam, giving the familiar diffuse effect of diffracted light. When there are two tunnels, the two-slit experiment, the interference pattern with its augmentation and nullification fringes appears (see Fig. 11).

A large scale experiment can be done to demonstrate this. If several different streams of BBs from one reservoir are aimed at several different streams of BBs from another reservoir, so that the streams collide at acute angles to each other, but each stream from each of the two reservoirs being at a little different angle from the others, as these multiple streams of BBs

Figure 11. Diffraction of light

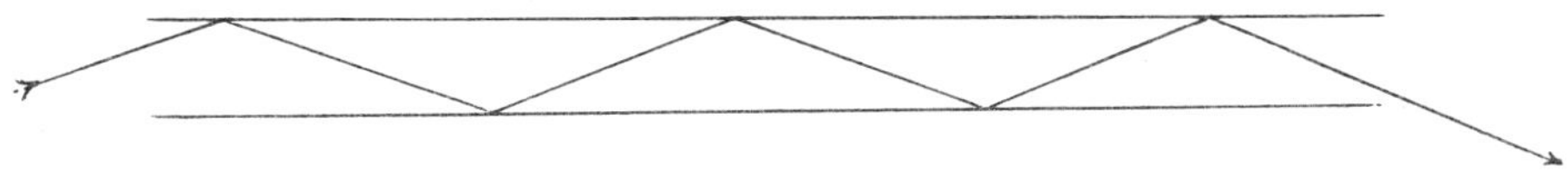

A. The theoretical path of a ray of light through a "slit"

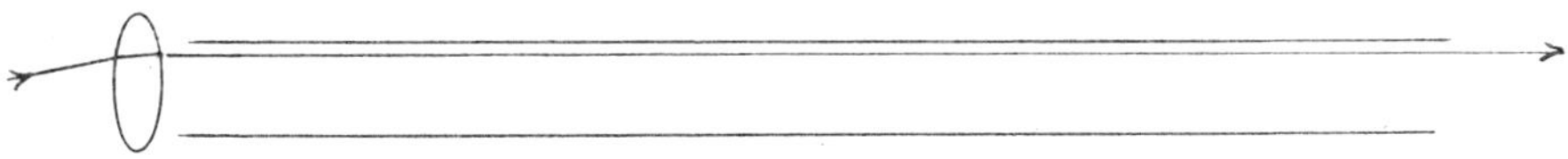

B. The theoretical path of a collimated ray of light through a "slit"

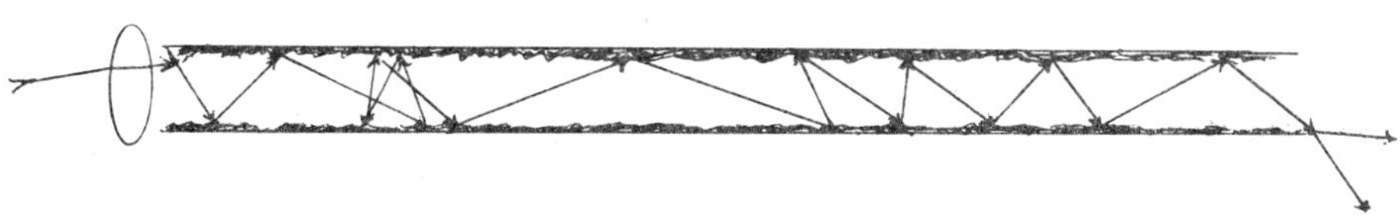

C. The probable actual path of a collimated ray of light through a "slit." The light "sees" jagged borders of the "slit," and the angle of incidence continues to equal the angle of reflection. Multiple collisions between various rays of light cause diffusion. Additionally, there may be a tangential swipe at the edge of exit.

(emags) roll down an inclined plane and are collected in an appropriate receiver which can accommodate one layer of BBs, it will be observed that the result is not the "normal distribution curve," but a pattern of highs and lows, remindful of light-interference patterns. As the quantum theory tells us, electromagnetic radiation behaves *as if* it is a wave form. But this is merely a statistical expectation of a large population of particles. In view of the Heisenberg uncertainty principle, will we ever be able to "track" single emags, no matter what the refinement of our instrumentation? Perhaps not, but let us leave the door ajar. And suppose we cannot; does that, therefore, mean that we should abandon the idea of the particulate nature of electromagnetic radiation? As mentioned earlier, we do not reject the molecular nature of water just because we see its waves. We should search for *truth,* and if truth is the particulate nature of electromagnetic radiation, that is what we should find. Truth is its own excuse (see Fig. 12).

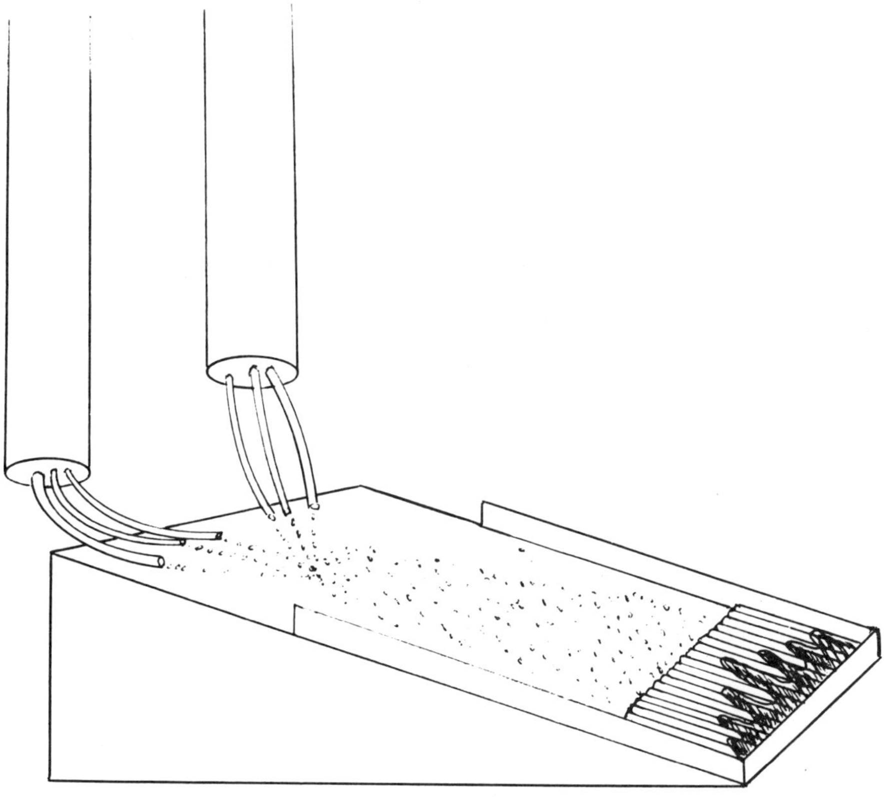

Figure 12. Interference pattern

7. *A Push or a Pull?*

Let us return to our postulates #6 and #7. How can we arrive at such postulates, considering our common experiences with magnetic, coulombic, and gravitational "force fields"? As mentioned earlier, this was a necessary sequel to the postulate of direct action, and the abandonment of the *vis viva* philosophy. Of course, it is not necessary to support a postulate, but it is neater to do so.

All of knowledge starts with empirical observation. Theories are then built upon such observations, and, if the theories have merit, they permit us to make predictions of as yet unobserved events. The ancients observed the sun and the planets and the "celestial sphere" revolve around the Earth. It was *obvious*. With the great ego of man, Ptolemy proposed the geocentric theory of the heavens, and it worked. Copernicus doubted, and developed the heliocentric theory (not original with him), and *it* worked. Kepler, with prodigous *empirical* observations and calculations, determined that Copernicus's belief in uniform circular motion was erroneous, and that the planetary orbits are ellipses. And then–another very important *empirical* observation which worked–the square of the periods equals the cube of the respective mean radii.

Granted, these empirical observations required later refinement by Newton, but bear in mind that they *were* empirical. Let us pursue Newton's great accomplishment. He proposed the law of universal gravitation. Note the word *proposed*. He demonstrated that the *force of gravitational "attraction"* was directly proportional to the product of the masses and inversely proportional to the square of the distance between two masses. Well, everybody knows that! But, how did he arrive at his proposal? By his own statement, Newton used Kepler's third law, in association with centripetal force, to achieve his formula:

$$F = \frac{Gm_1m_2}{d^2}$$

Note again—this is a development from an empirical observation. And since everybody knows that a body dropped from a height falls and is "attracted" to the Earth, Newton's law of universal gravitation became an attractive force, axiomatically.

Kepler had originally ascribed the motion of the planets around the sun to a *sweeping* force by the sun, but of course this was never accepted Looking at Newton's equation, again, we notice, as an empirical statement, that *there is a force whose effect is to cause the motions of the planets around the sun,* and as Kepler noted, in elliptical orbits. The equation does not state what the force is, nor that it is an attractive force. If we should suggest that the force is a push, rather than a pull, we have not changed the "truth" of the equation. We have simply given its *kind* a different interpretation. Could this be possible? The emag hypothesis says that it is.

The emag hypothesis also demands that magnetic and coulomb forces are pushes rather than pulls. But we see daily evidence that there are magnetic and coulombic "attractive" forces. Things are not always what they seem. Evacuate a closed tin can, and it collapses. Is this an attractive force? Push a rubber suction cup against a flat surface. Is this an attractive force? "But," you might say, "we see the air and we know it has very definite and measurable effects." Yes, this is a decided difference from magnetism. While we cannot know, in the absence of knowledge, that emags are there, nor that they have measurable effects, *if they are there,* cannot we ascribe these measurable magnetic forces to the push of emags? But if emags are there, and if their mass is almost zero, how could they have such large effects? It is merely a matter of credibility and the willingness to consider the possibility, for the number of emags is almost limitless. The attractive versus the repulsive effects of magnetism could be due to specific permeability, in one direction or the other, to the flow of ambient emags, as determined by the particular atomic alignment. Even the ocean tides would be explained, according to the emag hypothesis, as the resultant effect of the sea of emags from all of space, as the earth rotates on its axis.

Data were collected in an effort to associate the tides with the phases of the moon and sun. Since sunrise to sunset, throughout the year, has a minimal, gradual gradient, it was evident that the effect of the sun on the formation of tides was negligible. However, it was also evident that there was a cyclical relationship of the tides, which has been attributed, in the past, to the "attractive, gravitational effects of the moon." We are proposing another mechanism.

Since the emag hypothesis requires that the interrelationship of all bodies is due to the direct action or push of the ambient emag sea, it would appear that the first phase of the tides is a low tide, caused by the resultant emag push from space, on a fluid without resistance. Some hours later, the water, as in a basin, rolls back when the initial push has dissipated, causing the first high tide of the day. A repeat rolling of the water, back and forth, tending towards stability, creates the second low tide, followed by the second high tide, and then the cycle starts again the following day.

If the daily first push should not happen to be in phase with the returning high tide roll back of the previous day, some interference effects appear. This is, very likely, the usual state of affairs, for if it were not, each day's new tides, in phase with the previous day's tides, would inundate the Earth.

The interference effects described above, apparently, account for the fact that tides in various parts of the Earth do not fit neat patterns. Treatises on tides explain these effects by "other factors." Such treatises, too, display complex equations for the tides. These equations can be intimidating and seem to *document* their own authenticity. However, fitting equations to empirical observations does not tell *how,* only *how much.*

Another doubt needs to be raised. Newton's equation tells us that the "gravitational force" varies directly as the product of the masses and inversely as the square of the distance between two masses. The constant of proportionality, *G,* is assumed to be *constant.* In our local sphere of influence, i.e., the solar system, it may be. Perhaps it is even constant in our galaxy. But there is neither evidence nor any other reason to believe that *G* is constant throughout the universe. (See Chapter 10.)

The unorthodoxy of the push versus the pull may repel us, at first. But reason supports this position.

8. The Hydrogen Atom

At this juncture, let us return to the electron orbiting the nucleus of the hydrogen atom. Hydrogen, being the simplest atom, is useful as an example, though far from simple. Some questions come to mind. What keeps the hydrogen electron from flying out of its orbit? *Obviously,* the attractive coulomb force of the proton nucleus. What keeps the electron from spiralling into the proton? *Obviously,* the velocity of the electron, i.e., its *centrifugal force.* By virtue of our postulate #7 of the emag hypothesis, we have abandoned the concept of action at a distance, so we must seek another explanation–direct action. We shall accept, as a postulate, Newton's first law: a body in motion continues in its state of motion, unaltered, and in a straight line, unless acted upon by a force. Since orbital motion is not in a straight line, it follows that it is the result of a force. That force, according to the emag hypothesis, is a push. Consequently, it is proposed that the orbit of the electron is due to the resultant pushes of the sum of emags at any particular location. Why should the resultant push give orbital motion? Why not? Such a resultant push will give *some kind* of motion. Empirically, that proves to be orbital motion.

Now we are confronted with Bohr's hypothesis of different orbital energy states and "stationary orbits." This, of course, brings us to the question of spectral emission lines. Let us assume the basic first orbit to be the "stationary" state of Bohr. This is not associated with an emission line. Why not? Because it is an equilibrium state. Suppose the electron is kept in that orbit by the constant resultant rain of emags, and then something is added. Hertz and Franck added a stream of bombarding electrons. Suppose, in nature, a *gulf stream* of large aggregate emags strikes our electron. That electron, according to Newton's first law, will be accelerated and pushed into a "higher orbit." Since this higher energy orbit is an artificial one, and unstable, at the earliest opportunity, when the intruding force has passed, the original equilibrium state will be re-established. Why should the above motion be associated with an emission line? We must ask, "What is an emission line?" It is a beam of light of a particular frequency. As we have stated earlier, such a beam of light is a stream of "photons," or emag aggregates, having a particular frequency. That frequency is precisely the number of revolutions the orbital electron makes per second. It is proposed that every time an electron completes one revolution around its nucleus, in a single plane, it collides with the sea of emags, sending the emags recoiling, according to conservation of momentum, radially, in that plane. What the sensor receives is the same number of compressions and rarefactions of the emag sea, per second. The effect we observe in the spectrum is the sum effect of all the hydrogen atoms in the vicinity of the intruding force, as the velocity (and the frequency) of the orbiting electron goes into and out of different energy states. This explains why the Bohr stationary state is not associated with an emission line. In the stationary state, the electron is not buffeting the emag sea; it is being buffeted *by* the sea of emags, and this is the push that keeps it in orbit.

As is generally accepted, an accelerating body is constantly emitting radiation, and is losing energy. For this reason, it would be expected that an orbiting electron, for example, would quickly lose all its energy, and spiral into the nucleus. This expectation, however, is not the reality. Bohr tried to avoid this dilemma with the concept of the stationary state. The dilemma results from the attractive coulomb force's being balanced by the centrifugal effect of the rapidly moving electron, that is, a "built-in" energy system which ought to be depleting itself. According to the emag hypothesis, there is no need for a "built-in" energy system, since the electron is kept in orbit by the sea of emags, and the system does not deplete itself.

It is reasonable, and it has been assumed in the past, that a higher energy state is associated with a larger electron orbital radius, and that ionization occurs when the electron finally goes out of orbit. However, according to the emag hypothesis, a higher energy level of the orbiting electron is due to a *gulf stream* effect of larger emag aggregate fluxes, and the result is the forcing of the electron into a tighter orbit of smaller radius. Ionization occurs, then, when either by the thrust of the electron out of its tighter orbit, in a straight line, past the nucleus, or possibly even by crashing into the nucleus and ricocheting out.

According to the Bohr equations:

$$E_n = \frac{-2}{n^2} \frac{(\pi^2 e^4 m_e)}{h^2}$$

$$r_n = \frac{n^2}{4} \left(\frac{h^2}{\pi^2 e^2 m_e} \right)$$

Since E_n varies inversely as n^2, and r_n varies directly as n^2, E_n varies inversely as r_n, and this is what the emag hypothesis suggests. Since a higher energy state is associated with a greater frequency, this can be achieved either (1) by keeping the electron orbital velocity constant in a smaller orbital radius, or (2) by increasing the orbital velocity of the electron in a smaller orbital radius. Since, in the emag hypothesis, it is assumed that the higher energy state is due to an external force effect of larger emag aggregates buffeting the orbital electron into a tighter orbit, clearly, the momentum of the emag stream is greater, and, consequently, the momentum of the orbiting electron is increased. Since we are assuming the electron mass to be (fairly) constant, clearly the electron orbital velocity is increased, and the second alternative is the correct one.

9. *The "Photon" and the Velocity of Light*

Each energy orbit ought to be stable until changed. The "stationary state" is stable in the ambient sea of emags. A higher energy state is stable during the flow of the *gulf stream* of emag aggregates. However, at the moment the *gulf stream* stops buffeting the electron in the smaller, tighter, higher energy, higher velocity, higher frequency orbit, the stability of that orbit is interrupted, and, at that moment, the higher energy electron begins to buffet the ambient emag sea at its greater frequency, resulting in an emission line as it dissipates excess energy. That emission line will gradually broaden toward a lower frequency until the larger orbit of the next stable state is reached.

In view of the above, since we are dealing with different stable orbits, at different times, we ought to be able to use the "universal truth" of Kepler's third law, as Newton used it:

$$T^2 = r^3$$

We know the frequency of an emission line, whose reciprocal is the period, and the orbital radius is

$$T^{2/3} \text{ or } f^{-2/3}$$

for any stable orbit (see Table 1). Also, since, $V_e = 2\pi rf$, we can calculate V_e:

$$V_e = 2\pi f^{1/3}$$

at its moment of change from a higher energy orbit to a lower. Now, according to conservation of momentum,

$$m_{ph} V_{ph} = m_e V_e$$

We now have an equation with two known quantities and two unknown quantities. Planck's law, $E = hf$, tells us that the energy of a "photon" is directly proportional to the frequency. We have indicated the possibility of the greater energy of the greater frequency's being attributable either to a greater number of like and equal "photons" per unit time, or to a greater number of larger "photons." We shall assume the first alternative, first, for the sake of discussion. In so doing, we shall assume also, that the velocity of light, *c,* is constant. If *c* is constant, we can solve for m_{ph} (See Table 1) and note the range of "photon" masses for various frequencies. However, let us look at the second alternative, which we are inclined to favor for the following reason. According to $m_{ph}V_{ph} = m_eV_e$, since we know V_e and m_e, and if we assume m_{ph} to be fairly constant,

$$V_{ph} = \frac{m_e}{m_{ph}} \times V_e \qquad \text{and}$$

the velocity of light varies directly as the electron velocity, and is not constant.[1] As Table 1 shows, based on the above analysis, if the "photon" mass is considered constant, at 1.7790×10^{-33} Kgm, the velocity of light varies between 2.4779×10^8 m/sec. and 3.0141×10^8 m/sec., though the average is the accepted value of *c,* as it should be, because we are "juggling" averages. The variance of the velocity of light is totally unacceptable, and does not jibe with actual determinations. Consequently, the first alternative is untenable.

Now let us pursue the second alternative, i.e., that the greater energy of the greater frequency is due to the greater number of "photons" per unit time and *also* that the masses of these "photons" vary as the frequency.

According to Planck's Law: $E = hf$

According to Einstein: $E = mc^2$

Equating these two expressions: $hf = m_{ph}c^2$

We will not concern ourselves with the "rest mass" of the photon, because it is never at rest, and also because the "rest mass" of the "photon" is presumed to be zero, which is unacceptable. Consequently, since *c* is constant and *h* is constant,

$$m_{ph} = \frac{hf}{c^2} = \frac{6.6255 \times 10^{-34} \times f}{(2.9979 \times 10^8)^2}$$

1. In Chapter 5 we challenged Einstein's second postulate, as to the propriety of the velocity of light's being *unaffected* by the velocity of its source. We commented upon the asymptotic limit of particulate velocities. We concede that the velocity of light, itself, were it unaffected by the velocity of its source, might be *fairly* constant.

The mass of the "photon" varies as the frequency, by the constant

$$\frac{h}{c^2}$$

and the mass equals:

$$m_{ph} = 7.3720 \times 10^{-51} \times f \text{ Kgm}$$

According to our previous equation of conservation of momentum, in the hydrogen atom,

$$m_{ph} V_{ph} = m_e V_e \qquad \text{and}$$

$$m_{ph} = \frac{m_e}{V_{ph}} \times V_e$$

$$= 3.0385 \times 10^{-39} \times V_e \text{ Kgm}$$

Table 1 shows the variation of "photon" masses, according to the frequencies of the Balmer Series (see Column 9). With the number of emags calculated to be in the *assumed spherical electron*: 8.1520×10^6 (see p. 11), the known mass of the electron: 9.1091×10^{-31} Kgm; and the calculated mass of the emag: 1.1174×10^{-37} Kgm (see p. 13); knowing the range of masses of the "photon," we can calculate the number of emags per "photon."

Number of emags per "photon" $= \frac{\text{mass}_{ph}}{\text{mass}_{emag}}$ [2]

$$= \frac{1.4704 \times 10^{-33}}{1.1174 \times 10^{-37}} \text{ to } \frac{1.7886 \times 10^{-33}}{1.1174 \times 10^{-37}}$$

We have already indicated that the "photon" is not of constant size, nor is it believed that it is a discrete particle. It seems appropriate, in order to account for the wave-like manifestations of electromagnetic radiation, that what we have been accustomed to think of as a "photon" particle, is really a variable number of ambient emags, depending upon the electron velocity, which happen to lie in the path of a buffeting electron, in its orbit.

2. See footnote, page 10.

In one direction, in its orbit, the electron sets up a compression wave, and on the return trip, a rarefaction wave. This consideration explains, completely, the diffraction and interference effects of electromagnetic radiation, and also the Taylor experiment, where one "photon" aimed at two slits gives an interference pattern.

We have been speaking of the mass of the "photon." Inherent in the relativity concept, and Lorentz's *k,* is the notion that a "photon" have zero "rest mass." Otherwise, the mass of any moving object, including the lowly "photon," would approach infinity as its velocity approaches *c,* and of course, the velocity of the lowly "photon" reaches *c*. We have already demonstrated that "photons" have mass. Remember our postulate #3: mass is that which, on motion, has momentum and force. We can thus readily explain the force of electromagnetic radiation. Clearly, if "photons" have no mass, they can have no momentum nor force, *but they do*. Also, it is inconceivable that any body can be at complete rest, as Einstein's first postulate, correctly, states. Therefore, the introduction of the concept of "rest mass" appears to have been a *useful fiction*.

Getting back to the "photon," we shall turn to our interpretation of the photoelectric effect. As Planck observed, the energy of the "photon" is directly proportional to the frequency. In other words, the greater the frequency, the greater the number of "photons." The photoelectric effect is not the dislodgement of an electron by a single "photon," but rather by a wave of "photons" of a particular frequency, with sufficient energy to do the job.

Table 1

1	2	3	4	5	6	7	8	9	10	11_e
$m=$	$f=$ × $10^{14}/s$	$T=$ × $10^{-15}s$	"λ"= Å	$V_e=$ × $10^5 m/s$	$r=$ × $10^{-10}m$	$r=$ Å	$m_e/c=$ 3.0385 $\times 10^{-39}$	$m_{ph}=$ × 10^{-33}Kg	$m_e/av.m_{ph}$ = 512.03	"c"= $\frac{m_e \times V_e}{av. m_{ph}}$
3	4.5692	2.1886	6561.2	4.8394	1.6857	1.6857		1.4704		2.4779
4	6.1684	1.6212	4860.2	5.3486	1.3800	1.3800		1.6252		2.7386
5	6.9086	1.4475	4339.5	5.5545	1.2796	1.2796		1.6877		2.8441
6	7.3107	1.3678	4100.5	5.6602	1.2322	1.2322		1.7198		2.8982
7	7.5531	1.3240	3969.2	5.7221	1.2057	1.2057		1.7387		2.9299
8	7.7105	1.2969	3888.0	5.7616	1.1893	1.1893		1.7507		2.9501
9	7·8184	1.2790	3834.3	5.7883	1.1783	1.1783		1.7588		2.9638
10	7.8956	1.2665	3796.8	5.8073	1.1706	1.1706		1.7645		2.9735
11	7.9526	1.2574	3769.6	5.8212	1.1650	1.1650		1.7688		2.9806
12	7.9960	1.2506	3749.2	5.8318	1.1608	1.1608		1.7720		2.9860
13	8.0298	1.2454	3733.6	5.8400	1.1575	1.1575		1.7745		2.9902
14	8.0566	1.2412	3721.0	5.8465	1.1550	1.1550		1.7764		2.9936
15	8.0783	1.2379	3711.1	5.8518	1.1528	1.1528		1.7781		2.9963
16	8.0960	1.2318	3702.9	5.8560	1.1512	1.1512		1.7793		2.9984
17	8.1107	1.2329	3696.1	5.8596	1.1498	1·1498		1.7804		3.0003
18	8.1230	1.2311	3690.7	5.8625	1.1486	1.1486		1.7813		3.0018
19	8.1334	1.2295	3685.9	5.8650	1.1477	1.1477		1.7821		3.0030
20	8.1422	1.2282	3682.0	5.8671	1.1468	1.1468		1.7827		3.0041
21	8.1499	1.2270	3678.4	5.8690	1.1461	1.1461		1.7833		3.0051
22	8.1565	1.2260	3675.4	5.8706	1.1455	1.1455		1.7834		3.0059
23	8.1623	1.2251	3672.7	5.8720	1.1450	1.1450		1.7842		3.0066
24	8.1674	1.2244	3670.6	5.8732	1.1445	1.1445		1.7846		3.0072
25	8.1719	1.2237	3668.5	5.8743	1.1441	1.1441		1.7849		3.0078
26	8.1758	1.2231	3666.7	5.8752	1.1437	1.1437		1.7852		3.0083
27	8.1794	1.2226	3665.2	5.8761	1.1434	1.1434		1.7854		3.0087
28	8.1825	1.2221	3663.7	5.8768	1.1431	1.1431		1.7857		3.0091
29	8.1854	1.2217	3662.5	5.8775	1.1428	1.1428		1.7859		3.0094
30	8.1879	1.2213	3661.3	5.8781	1.1426	1.1426		1.7861		3.0098
31	8.1903	1.22126	3660.3	5.8782	1.1425	1.1425		1.7861		3.0098
100	8.2212	1.2164	3646.6	5.8861	1.1395	1.1395		1.7885		3.0138
125	8.2224	1.2162	3646.0	5.8863	1.1394	1.1394		1.7886		3.0140
150	8.2230	1.2161	3645.7	5.8865	1.1393	1.1393		1.7886		3.0141
200	8.2237	1.2160	3645.4	5.8866	1.1392	1.1392		1.7886		3.0141
AV.	at $m=$	125*						1.7790		2.9979

NOTES :

(2) $f = R(1/n^2 - 1/m^2)$ Balmer: $n = 2$
$R = 3.2898 \times 10^{15}$

$$m_{ph} = \frac{m_e}{V_{ph}} \times V_e = \frac{m_e}{c} \times V_e \qquad (9)$$

(6) $r = f^{-2/3}$ Kepler

$$V_{ph} = \frac{m_e}{m_{ph}} \times V_e$$

(4) "λ" $= c/f = cT$

(3) $T = 1/f$

$V_e = 2\pi rf$

(5) $= 2\pi f^{1/3}$

$m_e V_e = m_{ph} V_{ph} = m_{ph} c$, where $c = 2.9979 \times 10^8 m/s$, and

$m_e = 9.1091 \times 10^{-31}$ Kg

$$\text{"}c\text{"} = \frac{m_e}{av.m_{ph}} \times V_e \text{ , where "}c\text{"} \qquad (11)$$

is velocity of "photon" with constant average mass and is numerically equal to column 10 (512.03) $\times$ V_e.

$$\frac{m_e}{V_{ph}} = \frac{m_e}{C} = 3.0385 \times 10^{-39} \qquad (8)$$

$$\frac{m_e}{av.m_{ph}} = \frac{m_e}{1.7790 \times 10^{-33}} = 512.03 \qquad (10)$$

* Since the mass of the "photon" becomes essentially constant at $m = 125$ (column 9) and since r (column 7) varies by less than 0.0001 part in 10,000 after $m = 100$, $m = 125$ appears to be a good "end-point." The average m_{ph} is determined by the following method:

sum of m_{ph} from $m = 3$ to $m = 30$, inclusive (28 members)
= 49.1401

sum of m_{ph} from $m = 31$ to $m = 125$, inclusive (95 members)

$\frac{95}{2} \times$ (mass *ph* at $m = 31$ + mass *ph* at $m = 125$)
= 169.6820

Total = 218.8221

Divide by 123 members = 1.7790×10^{-33} Kgm

Similarly, "c" (column 11) was determined by adding

sum of "c" from $m = 3$ to $m = 30$, inclusive (28 members)
= 82.8083

sum of "c" from $m = 31$ to $m = 125$, inclusive (95 members)

$\frac{95}{2} \times$ ("c" at $m = 31$ + "c" at $m = 125$)

95 members = 285.9352

Total = 368.7435

Divide by 123 members = 2.9979×10^8m/sec.

10. The "Constancy" of G

In using the equation $T^2 = r^3$, in the previous section, as Kepler's third law, we are, obviously, considering the constant of proportionality to be 1. This is surely the case when we use as our units A.U. (astronomical units) and sidereal years. The question arises, of course, whether we are permitted to do so when we use as our units meters and seconds, as we are doing. Kepler used the units which were available to him, i.e., A.U.'s and sidereal years. The only distance dimension he had was the A.U., but he did have other dimensions of time. Using those other dimensions of time, his third law still was valid, except that K was no longer 1.

In setting up equations from proportionalities, we select, arbitrarily, those dimensions which allow a proportionality constant of 1, when possible. We even invent a unit, when necessary, in order that this should be the case. For example, we invent the dyne as that force which gives an acceleration of 1 cm/sec^2 to a mass of 1 gram. In the above considerations, we have elected, arbitrarily, in the manner of Kepler (because these are empirical data) to accept a proportionality constant of 1, since the data derived about orbital radii in the case of the hydrogen atom are completely in the same ball park as data from other sources. If K were truly significantly different from 1, this could not be the case. As a result of the above evaluation, we are led to other developments.

Newton's version of Kepler's third law is

$$(m_1 + m_2)\, T^2 = \frac{4\pi^2}{G}\,(r_1 + r_2)^3$$

In the case of the sun-Earth system, $m_1 + m_2$ (sun + Earth) is essentially m_1 (sun), and $r_1 + r_2$ (the distances of the sun and the Earth from their center of rotation) is essentially r_2. So our equation becomes

$$m_1 T^2 = \frac{4\pi^2}{G}\, r^3$$

And if we give m_1, the mass of the sun, unit value,

$$T^2 = \frac{4\pi^2}{G} r^3 \quad \text{and}$$

$$\frac{4\pi^2}{G} = K, \quad \text{the Kepler constant,}$$

which is 1 if our units are A.U. and sidereal years.

As can be seen, we have introduced a lot of unit values of 1, in order to arrive at our conclusion. At the same time, in the sun-Earth system,

$$G = 4\pi^2$$
and
$$G = \frac{39.4784 \text{ A.U.}^3}{\text{sun mass} - \text{Sid. yrs.}^2}$$

If we substitute equivalent Kgm-m-sec units, we get

$$G = \frac{39.4784 \times (1.495979 \times 10^{11} m)^3}{1.991 \times 10^{30} \text{Kgm} \times (3.155815 \times 10^7 \text{sec})^2}$$

$$= 6.6656 \times 10^{-11} m^3/\text{Kgm-sec}^2$$

Now let us proceed with our earlier assumption that for the atomic system, and ${}_1H^1$ in particular, since the radius of the orbital electron around the proton nucleus is in the same ball park using the Kepler formula with $K = 1$, as is found from other data, that K *is* essentially 1. Using Newton's formula for "universal" gravitation

$$F = \frac{Gm_p m_e}{r_e^2} \quad \text{and}$$

the formula for centripetal force

$$F = \frac{m_e V_e^2}{r_e}$$

and equating these two expressions,

$$\frac{Gm_p m_e}{r_e^2} = \frac{m_e V_e^2}{r_e} \quad \text{and}$$

multiplying both sides by r_e/m_e,

$$\frac{Gm_p}{r_e} = V_e^2$$

$$G = \frac{V_e^2 r_e}{m_p}$$

$m_p = 1.00054$ AMU $= 1.673024 \times 10^{-27}$Kgm (see p. 93)

r_e = from Lyman series (see Table 2, column 7, $m = 2$) $0.5404 \times 10^{-10}m$

= from Balmer series (see Table 2, column 7, $m = 10^3$) $1.0956 \times 10^{-10}m$

$V_e = 2\pi rf$ (Lyman $f_r = 2.5169 \times 10^{15}$/sec)
(Balmer $f_r = 0.87195 \times 10^{15}$/sec)

$= 2\pi \times 0.5404 \times 10^{-10}m \times 2.5169 \times 10^{15}$/sec (Lyman)
$= 2\pi \times 1.0956 \times 10^{-10}m \times 0.87195 \times 10^{15}$/sec (Balmer)

$V_e^2 = 7.30335 \times 10^{11} m^2/\text{sec}^2$ (Lyman)
$= 3.60286 \times 10^{11} m^2/\text{sec}^2$ (Balmer) and

$G = 2.3590 \times 10^{28} m^3/\text{Kgm-sec}^2$ (Lyman) and
$G = 2.3594 \times 10^{28} \text{m}^3/\text{Kgm-sec}^2$ (Balmer)

In other words, G is constant in the hydrogen system and it is constant in the solar system, but the two constants are not the same! We hinted earlier that G may not be a universal constant. The above exposition lends credence to that contention. Nor is this notion original with the present author. It would appear that there are different "gravitational" proportionality constants in different orbital planetary systems. In the hydrogen system, since we consider the mass of the proton to be constant, from

$$G = \frac{V_e^2 r_e}{m_p}$$

the product of V_e^2 and r_e must be constant.

We are still obliged to challenge the above statements. Fortunately, it is not difficult to do so. Let us return to our previous equality of

$$\frac{G m_p m_e}{r_e^2} = \frac{m_e V_e^2}{r_e}$$

$$G m_p = V_e^2 r_e$$

$$G = \frac{V_e^2 r_e}{m_p}$$

$$r_e = \frac{Gm_p}{V_e^2}$$

Let us solve for r_e using the recognized Kgm-m-sec value for G, the known mass of the proton and the previously used values for V_e^2, both for the Lyman and the Balmer series.

$$r_e = \frac{6.6656 \times 10^{-11} \times 1.673024 \times 10^{-27}}{7.30335 \times 10^{11}} \text{ (Lyman)}$$

$$= \frac{6.6656 \times 10^{-11} \times 1.673024 \times 10^{-27}}{3.60286 \times 10^{11}} \text{ (Balmer)}$$

$$r_e = 1.5269 \times 10^{-50}m = 1.5269 \times 10^{-40} \text{ Å !! (Lyman)}$$

$$= 3.0952 \times 10^{-49}m = 3.0952 \times 10^{-39} \text{ Å !! (Balmer)}$$

These results for the radius of the hydrogen electron are clearly untenable.

To pursue the matter even further, let us assume, merely for the sake of an approximation, that the velocity of the hydrogen electron is equal to the velocity of light, which is clearly not realistic. If we use as the velocity of light 3×10^8 m/sec, and as the radius of the orbital hydrogen electron $0.5 \times 10^{-10}m$, and as the mass of the proton 1.673×10^{-27} Kgm, and solve for G,

$$G = \frac{(3 \times 10^8)^2 \times 0.5 \times 10^{-10}}{1.673 \times 10^{-27}}$$

$$G = 2.6898 \times 10^{33} m^3/\text{Kgm-sec}^2$$

This value is nowhere near the accepted value of G.[1]

1. We have been using, as indicated, 1 as the constant of proportionality, in the Kepler formula, *because it works*. However, in the Newton formula,

$$F = \frac{Gm_1m_2}{r^2},$$

we are dealing with so-called *gravitational force*, with its *gravitational* constant of proportionality, while in the atomic system, we are still using the *gravitational* constant, when it might be contended that we ought to be using the *coulomb* constant, k $(1/4\pi\epsilon_o)$. These constants are, in a sense, "fudge factors," since they make the equations jibe with experimental, empirical data. In other words, when we use G for mass effects, things work out; and when we use k for "charge" effects, things work out. But the emag hypothesis denies the existence of charges. It contends that those phenomena which are presently interpreted as charge effects are, in reality, mass and negative mass effects. Whatever the interpretation, it requires k to make it work out. The emag hypothesis considers the G effect to be a distant effect and the k effect to be a near effect. Can we not, therefore, combine these two effects, mass (G) and "charge" (k), which is really mass and negative mass effect? Let us try:

$$F = \frac{Gkm_1m_2}{r^2} \qquad \text{and}$$

$$T^2 = Gkr^3$$

$$r^3 = \frac{T^2}{Gk} = f^{-2}G^{-1}k^{-1}$$

$$r = \sqrt[3]{f^{-2}G^{-1}k^{-1}} = f^{-2/3}\sqrt[3]{G^{-1}k^{-1}}$$

Calling the product Gk, S,

$$r = S^{-1/3}f^{-2/3}$$

$$S = Gk = (6.6732 \times 10^{-11}\frac{m^3}{\text{Kg-sec}^2})(8.9876 \times 10^9\frac{\text{newt} - m^2}{\text{coul}^2})$$

$$S = .59976\ m^6/\text{sec}^4\text{-coul}^2$$

$$S^{-1/3} = 1.1858$$

Interestingly, this value, $S^{-1/3}$, if used instead of 1, in the Kepler formula of pages 40 and 47, as the constant of proportionality, gives values which are entirely in the same ball-park. As a matter of fact,

$$1.1858 \times r_{rL} = 10^3 = 1.1858 \times .45117$$

$$= .53500\ \text{Å}$$

The Bohr radius is .52918 Å – within 1.1% of each other!

11. Real Orbital Frequencies

We have gotten a little ahead of the game. As was pointed out in the previous chapter, we had to make reference to data that have not yet been presented, further on in the text of this presentation. This was necessary in order that our data should jibe correctly with concepts, not yet presented. We had to hasten to get into the question of the presumed universal gravitational constant in order to explain why we were using 1 as the constant of proportionality in Kepler's third law. It would have been evident, even to the novice observer, that here was a question to be resolved. Now we must return to our previous discussion about the Bohr stationary state, emission lines, orbital frequencies, etc.

It has been assumed, in the previous considerations, that the frequencies we were dealing with were real frequencies. They are not. The Bohr atom has a so-called stationary state orbit which is presumed to be at a zero energy level. There is no reason to doubt that there is a *reference* orbit, which is stable in its own right by virtue of its existence in its own milieu. However, that reference orbit does have an electron in motion. That motion is an accelerated motion and an accelerated motion requires the existence of a real force. In the past, this accelerated motion, and associated force in the stable stationary state orbit, has been glossed over, and, in essence, overlooked. But it really exists, and the electron in that orbit has a real frequency. That frequency is due to the buffeting of the electron by the ambient emag sea, and is an entirely passive motion. Since the stationary state is not associated with an emission line, we do not know what its frequency is, at this point. However, we shall find it. We must bear in mind that the emission lines with which we are familiar are due to the change in frequency between two energy levels, and the frequency we *think* we are determining is really the sum of the recognized frequency *plus the basic frequency of the stationary state.*

In order to determine what the real orbital frequency of the hydrogen atom is, we shall proceed as follows: the first energy level of hydrogen, as determined by the energy required to give the first emission line, is 10.2 e.v.

$$\text{e.v.} = \frac{1.24 \times 10^4 \text{ e.v. Å}}{\lambda}, \text{ and} \qquad \lambda = \frac{1.24 \times 10^4 \text{ e.v. Å}}{\text{e.v.}}$$

$$\lambda = \frac{1.24 \times 10^4 \text{e.v. Å}}{10.2 \text{e.v.}} = 1216.9 \text{ Å}$$

Since the ionization energy of hydrogen is 13.6 e.v., the above calculation gives 912 Å as the limiting "wave length," just before ionization. These two "wave lengths" happen to be the limiting "wave lengths" of the Lyman series, and all the "action" in the Lyman series takes place in the interval of 3.4 e.v. We know, from the Lyman series, that the difference of 3.4 e.v. represents a difference in frequency of 3.2898×10^{15} minus 2.4674×10^{15}, or 0.8224×10^{15}. On the assumption (which is not true) that there is no basic frequency nor basic orbital radius (f_o and r_o, respectively), the difference in apparent radius, r_a, in the Lyman series is 0.0955Å.

However, according to the emag hypothesis, there are an f_o and an r_o, which we are trying to find. Let us make some exploratory calculations.

$$f_{L=2} \text{ is really } (f_o + f_{L=2}) = f_{r=2}$$

$$f_{L=1000} \text{ is really } (f_o + f_{L=1000}) = f_{r=1000}, \text{ etc.,}$$

where f_L refers to the frequency in the Lyman series when n is any integer whatever, which follows; f_o is the basic frequency; and f_r is the corresponding real frequency in the Lyman series, at the same "n".

Now, according to Kepler's third law,

$$(r_{rL=2})^3 = (f_o + f_{aL=2})^{-2} = \left[\frac{1}{(f_o + f_{aL=2})}\right]^2,$$

where $r_{rL=2}$ is the real radius when n is 2; in the Lyman series and $r_{aL=2}$ is the apparent radius, when the apparent frequency is $f_{aL=2}$. Consequently,

$$(r_{rL=2})^{3/2} = \frac{1}{(f_o + f_{aL=2})} = \frac{1}{\frac{1}{T_o} + \frac{1}{T_{aL=2}}}$$

$$= \frac{1}{\frac{T_{aL=2} + T_o}{T_o \times T_{aL=2}}}$$

$$= \frac{T_o \times T_{aL=2}}{T_o + T_{aL=2}}$$

where T_o and T_{aL} have the corresponding significance as the same subscripts in f_o and f_{aL}.

In Table 2, we assume that f_o might be approximately that of $f_{aP=6}$, and found the corresponding r_o. Then we determined f_r by adding f_o to f_a. Using this new f_r, we solved for the corresponding r_r in the series. This, of course, was based on the initial assumption, which was an outright guess.

Then we turned around, and made an entirely different set of assumptions. We assumed that since the ionization of hydrogen occurred at Lyman $m = 1000$, approximately, the change of radius near the ionization radius is so small that it might be reasonable to assume that the *difference* in the apparent radii between Lyman $m = 10$ and Lyman $m = 9$ would equal, approximately, the *difference* in the real radii between Lyman $m = 10$ and Lyman $m = 9$. Note, in the Table, that the range of apparent radii in the whole Lyman series is only 0.0955Å, and as the frequencies get greater (that is, approach the ionization frequency), the differences in successive radii get progressively smaller. So it seemed that the above was a reasonable assumption.

Now,

$$r = f^{-2/3}\,;\ r_a = f_a^{-2/3}\,;\ r_o = f_o^{-2/3}\,;\ r_r = f_r^{-2/3}$$

But

$$f_r = f_o + f_a, \text{ so}$$

$$r_r = (f_o + f_a)^{-2/3} \text{ and } r_{rL=9} = (f_o + f_{aL=9})^{-2/3} \text{ and}$$

$$r_{rL=10} = (f_o + f_{aL=10})^{-2/3}$$

Using our assumption that $r_{rL=9} - r_{rL=10} = r_{aL=9} - r_{aL=10}$,

$$r_{aL=9} - r_{aL=10} = (f_o + f_{aL=9})^{-2/3} - (f_o + f_{aL=10})^{-2/3} \text{ or}$$

$$r_{aL=9} - r_{aL=10} = (f_{aL=9} + f_o)^{-2/3} - (f_{aL=10} + f_o)^{-2/3}$$

We assume now, that $f_o \ll f_{aL}$. If this were not so, small energy changes in the lesser series (i.e. Brackett and Pfund) would not cause material changes in frequencies, but they do. So, assuming that $f_o \ll f_{aL}$, the last equation is the difference between two binomial expansions, each of which is a converging infinite series.

$$r_{aL=9} - r_{aL=10} = f_{aL=9}^{-2/3} \times \left(1 + \frac{f_o}{f_{aL=9}}\right)^{-2/3} - f_{aL=10}^{-2/3} \times \left(1 + \frac{f_o}{f_{aL=10}}\right)^{-2/3}$$

and, calling $\frac{f_o}{f_{aL=9}}$ K_1 and $\frac{f_o}{f_{aL=10}}$ K_2

$$r_{aL=9} - r_{aL=10} = \Bigg[1^{-2/3} + \frac{(-2/3)\,(1^{-5/3})K_1}{1!} + \frac{(-2/3)\,(-5/3)\,(1^{-8/3})K_1^2}{2!} +$$

$$\frac{(-2/3)\,(-5/3)\,(-8/3)\,(1^{-11/3})\,K_1^3}{3!} +$$

$$\frac{(-2/3)\,(-5/3)\,(-8/3)\,(-11/3)\,(1^{-14/3})\,K_1^4}{4!} + \ldots\Bigg]\; f_{aL=9}{}^{-2/3}$$

$$-\Bigg[1^{-2/3} + \frac{(-2/3)\,(1^{-5/3})\,K_2}{1!} +$$

$$\frac{(-2/3)\,(-5/3)\,(1^{-8/3})\,K_2^2}{2!} +$$

$$\frac{(-2/3)\,(-5/3)\,(-8/3)\,(1^{-11/3})\,K_2^3}{3!} +$$

$$\frac{(-2/3)\,(-5/3)\,(-8/3)\,(-11/3)\,(1^{-14/3})\,K_2^4}{4!} + \ldots\Bigg]\; f_{aL=10}{}^{-2/3}$$

$$r_{aL=9} - r_{aL=10} = r_{aL=9}\Bigg[1 - \frac{.66667f_o}{f_{aL=9}} + \frac{.55556f_o^2}{(f_{aL=9})^2} - \frac{.49383f_o^3}{(f_{aL=9})^3} +$$

$$\frac{.45267f_o^4}{(f_{aL=9})^4} - \ldots\Bigg]$$

$$- r_{aL=10}\Bigg[1 - \frac{.66667f_o}{f_{aL=10}} + \frac{.55556f_o^2}{(f_{aL=10})^2} - \frac{.49383f_o^3}{(f_{aL=10})^3} +$$

$$\frac{.45267f_o^4}{(f_{aL=10})^4} - \ldots\Bigg]$$

We will use as $f_{aL=9}$: 3.2492×10^{15} $(r_{aL=9} = .4558\ \text{Å})$ and

$f_{aL=10}$: 3.2569×10^{15} $(r_{aL=10} = .4551\ \text{Å})$

Therefore, $.0007\ \text{Å} = (f_{aL=9} + f_o)^{-2/3} - (f_{aL=10} + f_o)^{-2/3}$ and

$.0007\,\text{Å} = (3.2492 \times 10^{15} + f_o)^{-2/3} - (3.2569 \times 10^{15} + f_o)^{-2/3}$

We try, empirically, various values of f_o.

When we try $f_o = .0495 \times 10^{15}$, our equation becomes

$$.0007Å = [(3.2492 + .0495) \times 10^{15}]^{-2/3} - [(3.2569 + .0495) \times 10^{15}]^{-2/3}$$

$$= (3.2987 \times 10^{15})^{-2/3} - (3.3064 \times 10^{15})^{-2/3} \text{ and}$$

$$.0007Å = .0007\,Å$$

So $f_o = .0495 \times 10^{15}$ fits the equation, and
$r_o = 7.4176\,Å$

which is very close to our first assumption—as a guess.

There is another approach:

Since $r = f^{-2/3}$, $r^{3/2} = f^{-1}$

$$(r_{rL=9})^{3/2} - (r_{rL=10})^{3/2} = (f_{rL=9})^{-1} - (f_{rL=10})^{-1}$$

$$= (f_{aL=9} + f_o)^{-1} - (f_{aL=10} + f_o)^{-1}$$

$$= \frac{1}{(f_{aL=9} + f_o)} - \frac{1}{(f_{aL=10} + f_o)}$$

$$= \frac{f_{aL=10} + f_o - f_{aL=9} - f_o}{f_o^2 + f_o\,(f_{aL=9} + f_{aL=10}) + (f_{aL=9})\,(f_{aL=10})}$$

and $$(r_{rL=9})^{3/2} - (r_{rL=10})^{3/2} = \frac{f_{aL=10} - f_{aL=9}}{f_o^2 + f_o\,(f_{aL=9} + f_{aL=10}) + (f_{aL=9})\,(f_{aL=10})}$$

If $f_o = .0495 \times 10^{15}$ and $r_o = 7.4176\,Å$, then

$$r_{rL=9} = (f_{aL=9} + f_o)^{-2/3} = [(3.2492 + .0495) \times 10^{15}]^{-2/3}$$

$$= (3.2987 \times 10^{15})^{-2/3}$$

$$= .4315\,Å$$

Similarly, $r_{rL=10} = .4506\,Å$ and

$$r_{rL=9} - r_{rL=10} = .0007\,Å$$

From the equation with the quadratic form in the denominator,

$$(r_{rL=9})^{3/2} - (r_{rL=10})^{3/2} = \frac{3.2569 \times 10^{15} - 3.2492 \times 10^{15}}{(.0495 \times 10^{15})^2 + (.0495)\,(6.5061)\,(10^{30}) + (3.2569)\,(3.2492)\,(10^{30})}$$

$$(.4513 \times 10^{-10})^{3/2} - (.4506 \times 10^{-10})^{3/2} = \frac{.0077 \times 10^{15}}{10.9764 \times 10^{30}}$$

$$.0007 \times 10^{-15} = .0007 \times 10^{-15}$$

and this is an identity, showing that, using these equations and the derived values of f_o and r_o, the equations work.

Now let us see if these values of f_o and r_o fit the other series.

$$(r_{rB=24})^{3/2} - (r_{rB=25})^{3/2} = \frac{.81719 \times 10^{15} - .81674 \times 10^{15}}{(.0495 \times 10^{15})^2 + (.0495)(1.63393)(10^{30}) + (.81719)(.81674)(10^{30})}$$

$$(1.1005 \times 10^{-10})^{3/2} - (1.1001 \times 10^{-10})^{3/2} = \frac{.00045 \times 10^{15}}{7.5076 \times 10^{29}}$$

$$.00063 \times 10^{-15} = .00060 \times 10^{-15}$$

The result is satisfactory for the Balmer series.

$$(r_{rP=47})^{3/2} - (r_{rP=48})^{3/2} = \frac{.13016 \times 10^{15} - .13010 \times 10^{15}}{(.0495 \times 10^{15})^2 + (.0495)(.26026)(10^{30}) + (.13016)(.13010)(10^{30})}$$

$$(3.1407 \times 10^{-10})^{3/2} - (3.1414 \times 10^{-10})^{3/2} = \frac{.00006 \times 10^{15}}{1.9384 \times 10^{28}}$$

$$.00185 \times 10^{-15} = .0031 \times 10^{-15}$$

The result is tolerable for the Pfund series.

In trying to fit f_o into the tables for each series, we select, in each series that $f_{am+1} - f_m$, such that $r_{rm} - r_{rm+1}$ is approximately equal to $r_{am} - r_{am+1}$. As we have seen, when f_o is $.0495 \times 10^{15}$, the results are satisfactory.

The question arises–can we zero in on a better value of f_o? Using the same empirical approach as just used, but using radii calculations to five significant figures instead of four, we find that a value of f_o of 0.01000×10^{15}/sec. gives "sharper" results. Table 2 shows these data. Additionally, following the same analysis, with $f_o = 0.01000 \times 10^{15}$, gives the following results:

$$(r_{r(m-1)})^{3/2} - (r_{r(m)})^{3/2} = \frac{f_{a(m)} - f_{a(m-1)}}{f_o^2 + f_o(f_{a(m-1)} + f_{a(m)}) + (f_{a(m-1)})(f_{a(m)})}$$

Lyman series: $m = 10$ $m - 1 = 9$

$$f_{aL=10} = 3.2569 \times 10^{15} \qquad f_{aL=9} = 3.2492 \times 10^{15}$$

$$r_{rL=10} = .45420\,\text{Å} \qquad r_{rL=9} = .45491\,\text{Å}$$

$$(r_{rL=9})^{3/2} - (r_{rL=10})^{3/2} = \frac{f_{aL=10} - f_{aL=9}}{f_o^2 + f_o(f_{aL=9} + f_{aL=10}) + (f_{aL=9})(f_{aL=10})}$$

$$3.0682 \times 10^{-16} - 3.0610 \times 10^{-16} = \frac{.0077 \times 10^{15}}{1.0000 \times 10^{26} + 6.5061 \times 10^{28} + 1.0582 \times 10^{31}}$$

$$7.2 \times 10^{-19} = \frac{.0077 \times 10^{15}}{1.0647 \times 10^{31}}$$

$7.2 \times 10^{-19}m = 7.2 \times 10^{-19}m$, an identity.

Balmer series: $m = 16$ $m - 1 = 15$

$$f_{aB=16} = .80960 \times 10^{15} \qquad f_{aB=15} = .80783 \times 10^{15}$$

$$r_{rB=16} = 1.1418\,\text{Å} \qquad r_{rB=15} = 1.1435\,\text{Å}$$

$$(r_{rB=16})^{3/2} - (r_{rB=15})^{3/2} = \frac{f_{aB=16} - f_{aB=15}}{f_o^2 + f_o(f_{aB=15} + f_{aB=16}) + (f_{aB=15})(f_{aB=16})}$$

$$1.2228 \times 10^{-15} - 1.2201 \times 10^{-15} = \frac{.00177 \times 10^{15}}{1.0000 \times 10^{26} + 1.6174 \times 10^{28} + 6.5402 \times 10^{29}}$$

$$2.7 \times 10^{-18} = \frac{.00177 \times 10^{15}}{6.7029 \times 10^{29}}$$

$2.7 \times 10^{-18}m = 2.64 \times 10^{-18}m$; very adequate.

Pfund series: $m = 48$ $m - 1 = 47$

$$f_{aP=48} = .13106 \times 10^{15} \qquad f_{aP=47} = .13010 \times 10^{15}$$

$$r_{rP=48} = 3.7061\,\text{Å} \qquad r_{rP=47} = 3.7072\,\text{Å}$$

$$(r_{rP=47})^{3/2} - (r_{rP=48})^{3/2} = \frac{f_{aP=48} - f_{aP=47}}{f_o^2 + f_o(f_{aP=47} + f_{aP=48}) + (f_{aP=47})(f_{aP=48})}$$

$$7.1379 \times 10^{-15} - 7.1347 \times 10^{-15} = \frac{.00006 \times 10^{15}}{1.0000 \times 10^{26} + 2.6026 \times 10^{27} + 1.6934 \times 10^{28}}$$

$$3.2 \times 10^{-18} = \frac{.00006 \times 10^{15}}{1.9637 \times 10^{28}}$$

$$3.2 \times 10^{-18} m = 3.06 \times 10^{-18} m \text{ , very adequate.}$$

The following diagrams summarize the above:

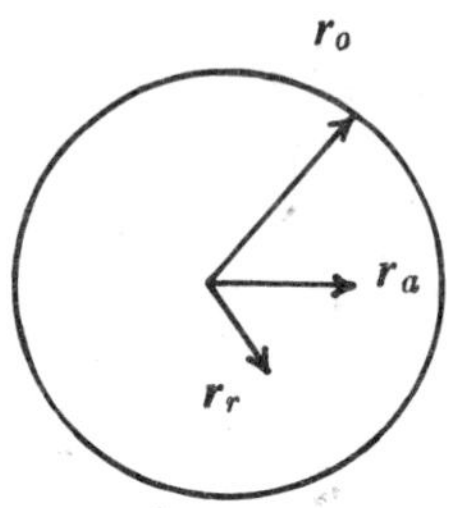

if: $f_o = .4500 \times 10^{15}$; $r_o = 1.7029$Å;
$f_a = 2.4674 \times 10^{15}$; $r_a = .5476$Å
Then: $f_r = 2.7174 \times 10^{15}$; and $r_r = .4898$Å

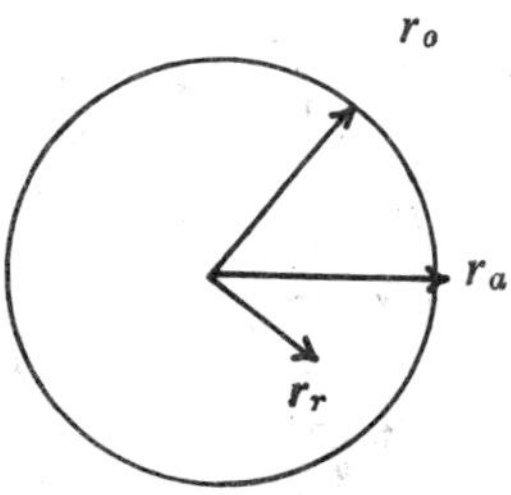

if: $f_o = 2.5000 \times 10^{15}$; $r_o = .5429$Å
$f_a = 2.4674 \times 10^{15}$; $r_a = .5476$Å
Then: $f_r = 4.9674 \times 10^{15}$; and $r_r = .3435$Å

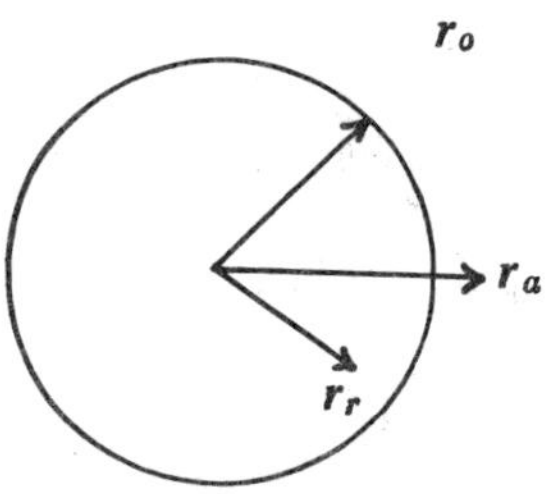

if: $f_o = 9.0000 \times 10^{15}$; $r_o = .2311$Å
$f_a = 2.4674 \times 10^{15}$; $r_a = .5477$Å
Then: $f_r = 11.467 \times 10^{15}$; and $r_r = .1966$Å

Thus r_r is always less than r_a; r_r is always less than r_o; r_a can be greater or less than r_o.

$$\text{If } T_o \gg T_a \text{ or } f_o \ll f_a, \text{ then } (r_r)^{3/2} \cong \frac{T_o T_a}{T_o} \cong T_a \cong \frac{1}{f_a} \text{ and}$$

$$r_r \cong f_a^{-2/3} \cong r_a$$

$$\text{If } T_o \ll T_a \text{ or } f_o \gg f_a, \text{ then } (r_r)^{3/2} \cong \frac{T_o T_a}{T_a} \cong T_o \cong \frac{1}{f_o} \text{ and}$$

$$r_r \cong f_o^{-2/3} \cong r_o$$

If $T_o \cong T_a$ or $f_o \cong f_a$, then $(r_r)^{3/2} \cong \frac{T_a T_a}{2T_a} \cong \frac{T_a}{2} \cong \frac{1}{2f_a}$ and

$$r_r \cong (2f_a)^{-2/3}$$

Table 2

NOTES:

Table 2 looks ominous, and is quite lengthy, but only for purposes of documentation. The following notes are of importance:

Each section of the table shows the following data, derived from the Rydberg formula.

$$\text{Frequency} = 3.2898 \times 10^{15} (1/n^2 - 1/m^2)$$

Column #1 shows which series we are dealing with, i.e., Lyman ($n = 1$), Balmer ($n = 2$), Paschen ($n = 3$), Brackett ($n = 4$), and Pfund ($n = 5$), with m as indicated.

Column #2 indicates, in each section of the Table, what the presumed f_o is. This f_o (column #2), added to f_a (column #4), gives the presumed *real* f, or f_r, which is in Column #6

Column #3 gives the presumed *base* radius, r_o, for the presumed base frequency, f_o

Column #4 shows the corresponding frequency: f_a.

Column #5 indicates what the corresponding radius for any given f_a is, as derived from Kepler's third law.

$$r^3 = T^2 = f^{-2}$$

$$r = f^{-2/3}$$

The subscript *a* refers to *apparent,* which is the value as derived from the Rydberg formula and from experimental spectroscopy. According to the emag hypothesis, the apparent frequency, f_a, is really the frequency superimposed on the frequency of the "stationary state," f_o.

Column #8 gives the range of r_a in each series, for any particular f_a.

Column #9 gives the range of r_r in each series for any particular f_r.

Column #10 shows the difference in consecutive apparent radii, $r_{a(m)} - r_{a(m+1)}$ except where not consecutive.

Column #11 shows the difference in consecutive presumed real radii, $r_{r(m)} - r_{r(m+1)}$ except where not consecutive.

1	2	3	4	5	6	7	8	9	10	11
Lyman: $m=$	$f_o=$ .0495 $\times 10^{15}$	$r_o=$ 7.4176 Å	$f_a=$ $\times 10^{15}$	$r_a=$ Å	$f_r=$ (2) +(4) = $\times 10^{15}$	$r_r=$ Å	Range $r_a=$ Å	Range $r_r=$ Å	r_a: $m-(m+1)$	r_r: $m-(m+1)$
2			2.4674	.5476	2.5169	.5404	.0955	.0928		
3			2.9243	.4890	2.9738	.4836			.0586	.0568
4			3.0842	.4720	3.1337	.4670			·0170	.0166
5			3.1582	.4646	3.2077	.4598			.0074	.0072
6			3.1984	.4606	3.2479	.4560			.0040	.0038
7			3.2227	.4583	3.2722	.4537			.0023	.0023
8			3.2384	.4568	3.2879	.4522			.0015	.0015
9			3.2492	.4558	3.2987	.4513			.0010	.0009
10			3.2569	.4551	3.3064	.4506			.0007	.0007
18			3.2796	.4530	3.3291	.4485			.0021	.0021
10^3			3.2898	.4521	3.3393	.4476			.0009	.0009
Balmer: $m=$										
3			.45692	1.6857	.50642	1.5740	.5465	.4784		
4			.61684	1.3800	.66634	1.3108			.3057	.2632
5			.69086	1.2796	.74036	1.2219			.1004	.0889
6			.73107	1.2322	.78057	1.1796			.0474	.0423
7			.75531	1.2057	.80481	1.1558			.0265	.0238
8			.77105	1·1893	.82055	1.1409			.0164	.0149
24			.81674	1.1445	.86624	1.1005			.0448	.0404
25			.81719	1.1441	.86669	1.1001			.0004	.0004
10^3			.82245	1.1392	.87195	1.0956			.0049	.0045

$f = R(1/n^2 - 1/m^2)$; $f = 3.2898 \times 10^{15} (1/n^2 - 1/m^2)$

1	2	3	4	5	6	7	8	9	10	11
Paschen: $m=$	$f_o=$ .0495 $\times 10^{15}$	$r_o=$ 7.4176 Å	$f_a=$ $\times 10^{15}$	$r_a=$ Å	$f_r=$ (2) +(4) = $\times 10^{15}$	$r_r=$ Å	Range $r_a=$Å	Range $r_r=$Å	r_a: $m-$ $(m+1)$	r_r: $m-$ $(m+1)$
4			.15992	3.3942	.20942	2.8357	1.4381	1.0384		
5			.23394	2.6339	.28344	2.3175			.7603	.5182
6			.27415	2.3696	.32365	2.1214			.2643	.1961
7			.29839	2.2395	.34789	2.0216			.1301	.0988
8			.31413	2.1640	.36363	1.9629			.0755	.0587
25			.36026	1.9751	.40976	1.8126			.1889	.1503
26			.36066	1.9736	.41016	1.8115			.0015	.0011
10^3			.36553	1.9561	.41503	1.7973			.0175	.0142
Brackett: $m=$										
5			.07402	5.6724	.12352	4.0319	2.8018	1.5458		
6			.11423	4.2476	.16373	3.3413			1.4248	.6906
7			.13847	3.7362	.18797	3.0475			.5114	.2938
8			.15421	3.4774	.20371	2.8884			.2588	.1591
30			.20196	2.9051	.25146	2.5101			.5723	.3783
31			.20219	2.9029	.25169	2.5086			.0022	.0015
10^3			.20561	2.8706	.25511	2.4861			.0323	.0225
Pfund: $m=$										
6			.04021	8.5201	.08971	4.9901	4.6548	1.8636		
7			.06445	6.2209	.11395	4.2546			2.2992	.7345
8			.08019	5.3776	.12969	3.9030			.8433	.3516
47			.13010	3.8948	.17960	3.1414			1.4828	.7616
48			.13016	3.8936	.17966	3.1407			.0012	.0007
10^3			.13159	3.8653	.18089	3.1265			.0283	.0142

1	2	3	4	5	6	7	8	9	10	11
Lyman: $m=$	$f_o=$ .0740 $\times 10^{15}$	$r_o=$ 5.6734 Å	$f_a=$ $\times 10^{15}$	$r_a=$ Å	$f_r=$ (2) +(4) = $\times 10^{15}$	$r_r=$ Å	Range $r_a=$ Å	Range $r_r=$ Å	r_a: $m-(m+1)$	r_r: $m-(m+1)$
2			2.4674	.5476	2.5414	.5370	.0955	.0916		
3			2.9243	.4890	2.9983	.4809			.0586	.0561
4			3.0842	.4720	3.1582	.4646			.0170	.0163
5			3.1582	.4646	3.2322	.4574			.0074	.0072
6			3.1984	.4606	3.2724	.4537			.0040	.0037
7			3.2227	.4583	3.2967	.4514			.0023	.0023
8			3.2384	.4568	3.3124	.4500			.0015	.0014
9			3.2492	.4558	3.3232	.4490			.0010	.0010
10			3.2569	.4551	3.3309	.4484			.0007	.0006
18			3.2796	.4530	3.3536	.4463			.0021	.0021
10^3			3.2898	.4521	3.3638	.4454			.0009	.0009
Balmer: $m=$										
3			.45692	1.6857	.53092	1.5252	.5465	.4496		
4			.61684	1.3800	.69084	1.2796			.3057	.2456
5			.69086	1.2796	.76486	1.1957			.1004	.0839
6			.73107	1.2322	.80507	1.1555			.0474	.0402
7			.75531	1.2057	.82931	1.1329			.0265	.0226
8			.77105	1.1893	.84505	1.1188			.0164	.0141
24			.81674	1.1445	.89074	1.0802			.0448	.0386
25			.81719	1.1441	.89119	1.0798			.0004	.0004
10^3			.82245	1.1392	.89645	1.0756			.0049	.0042

1	2	3	4	5	6	7	8	9	10	11
Paschen: $m=$	$f_o=$ .0740 $\times 10^{15}$	$r_o=$ 5.6734 Å	$f_a=$ $\times 10^{15}$	$r_a=$ Å	$f_r=$ (2) +(4) = $\times 10^{15}$	$r_r=$ Å	Range $r_a=$ Å	Range $r_r=$ Å	r_a: $m-(m+1)$	r_r: $m-(m+1)$
4			.15992	3.3942	.23392	2.6340	1.4381	.9042		
5			.23394	2.6339	.30794	2.1929			.7603	.4411
6			.27415	2.3696	.34815	2.0206			.2643	.1723
7			.29839	2.2395	.37239	1.9186			.1301	.1020
8			.31413	2.1640	.38813	1.8794			.0755	.0392
25			.36026	1.9751	.43426	1.7438			.1889	.1356
26			.36066	1.9736	.43466	1.7427			.0015	.0011
10^3			.36553	1.9561	.43953	1.7298			.0175	.0129
Brackett: $m=$										
5			.07402	5.6724	.14802	3.5737	2.8018	1.2351		
6			.11423	4.2476	.18823	3.0447			1.4248	.5290
7			.13847	3.7362	.21247	2.8085			.5114	.2362
8			.15421	3.4774	.22821	2.6778			.2588	.1307
30			.20196	2.9051	.27596	2.3592			.5723	.3186
31			.20219	2.9029	.27619	2.3579			.0022	.0013
10^3			.20561	2.8706	.27961	2.3386			.0323	.0193
Pfund: $m=$										
6			.04021	8.5201	.11421	4.2481	4.6548	1.3773		
7			.06445	6.2209	.13845	3.7366			2.2992	.5115
8			.08019	5.3776	.15419	3.4777			.8433	.2589
47			.13010	3.8948	.20410	2.8847			1.4828	.5930
48			.13016	3.8936	.20416	2.8842			.0012	.0005
10^3			.13159	3.8653	.20559	2.8708			.0283	.0134

1	2	3	4	5	6	7	8	9	10	11
Lyman: $m=$	$f_o=$ .1500 $\times 10^{15}$	$r_o=$ 3.5422 Å	$f_a=$ $\times 10^{15}$	$r_a=$ Å	$f_r=$ (2) +(4) = $\times 10^{15}$	$r_r=$ Å	Range $r_a=$ Å	Range $r_r=$ Å	r_a: $m-(m+1)$	r_r: $m-(m+1)$
2			2.4674	.5476	2.6174	.5263	.0955	.0877		
3			2.9243	.4890	2.0743	.4730			.0586	.0533
4			3.0842	.4720	3.2342	.4572			.0170	.0158
5			3.1582	.4646	3.3082	.4504			.0074	.0068
6			3.1984	.4606	3.3484	.4468			.0040	.0036
7			3.2227	.4583	3.3727	.4446			.0023	.0022
8			3.2384	.4568	3.3884	.4433			.0015	.0013
9			3.2492	.4558	3.3992	.4423			.0010	.0010
10			3.2569	.4551	3.4069	.4417			.0007	.0006
18			3.2796	.4530	3.4296	.4397			.0021	.0020
10^3			3.2898	.4521	3.4398	.4388			.0009	.0009
Balmer: $m=$										
3			.45692	1.6857	.60692	1.3950	.5465	.3762		
4			.61684	1.3800	.76684	1.1936			.3057	.2014
5			.69086	1.2796	.84086	1.1225			.1004	.0711
6			.73107	1.2322	.88107	1.0881			.0474	.0344
7			.75531	1.2057	.90531	1.0686			.0265	.0195
8			.77105	1.1893	.92105	1.0564			.0164	.0122
24			.81674	1.1445	.96674	1.0228			.0448	.0336
25			.81719	1.1441	.96719	1.0225			.0004	.0003
10^3			.82245	1.1392	.97245	1.0188			.0049	.0037

1	2	3	4	5	6	7	8	9	10	11
Paschen: $m=$	$f_o=$ .1500 $\times10^{15}$	$r_o=$ 3.5422 Å	$f_a=$ $\times10^{15}$	$r_a=$ Å	$f_r=$ (2) +(4) = $\times10^{15}$	$r_r=$ Å	Range $r_a=$ Å	Range $r_r=$ Å	r_a: $m-(m+1)$	r_r: $m-(m+1)$
4			.15992	3.3942	.30992	2.1836	1.4381	.6282		
5			.23394	2.6339	.38394	1.8930			.7603	.2906
6			.27415	2.3696	.42415	1.7714			.2643	.1216
7			.29839	2.2395	.44839	1.7070			.1301	.0644
8			.31413	2.1640	.46413	1.6682			.0755	.0388
25			.36026	1.9751	.51026	1.5660			.1889	.1022
26			.36066	1.9736	.51066	1.5652			.0015	.0008
10^3			.36553	1.9561	.51553	1.5554			.0175	.0098
Brackett: $m=$										
5			.07402	5.6724	.22402	2.7111	2.8018	.7188		
6			.11423	4.2476	.26423	2.4285			1.4248	.2826
7			.13847	3.7362	.28847	2.2905			.5114	.1380
8			.15421	3.4774	.30421	2.2108			.2588	.0797
30			.20196	2.9051	.35196	2.0060			.5723	.2048
31			.20219	2.9029	.35219	2.0052			.0022	.0008
10^3			.20561	2.8706	.35561	1.9923			.0323	.0129
Pfund: $m=$										
6			.04021	8.5201	.19021	3.0235	4.6548	.6958		
7			.06445	6.2209	.21445	2.7911			2.2992	.2324
8			.08019	5.3776	.23019	2.6624			.8433	.1257
47			.13010	3.8948	.28010	2.3359			1.4828	.3265
48			.13016	3.8936	.28016	2.3356			.0012	.0003
10^3			.13159	3.8653	.28159	2.3277			.0283	.0079

1	2	3	4	5	6	7	8	9	10	11
Lyman: $m=$	$f_o=$ .4500 $\times 10^{15}$	$r_o=$ 1.7029 Å	$f_a=$ $\times 10^{15}$	$r_a=$ Å	$f_r=$ (2) +(4) = $\times 10^{15}$	$r_r=$ Å	Range $r_a=$ Å	Range $r_r=$ Å	r_a: $m-$ $(m+1)$	r_r: $m-$ $(m+1)$
2			2.4674	.5476	2.9174	.4898	.0955	.0748		
3			2.9243	.4890	3.3743	.4445			.0586	.0453
4			3.0842	.4720	3.5342	.4310			.0170	.0135
5			3.1582	.4646	3.6082	.4251			.0074	.0059
6			3.1984	.4606	3.6484	.4220			.0040	.0031
7			3.2227	.4583	3.6727	.4201			.0023	.0019
8			3.2384	.4568	3.6884	.4189			.0015	.0012
9			3.2492	.4558	3.6992	.4181			.0010	.0008
10			3.2569	.4551	3.7069	.4175			.0007	.0006
18			3.2796	.4530	3.7296	.4158			.0021	.0017
10^3			3.2898	.4521	3.7398	.4150			.0009	.0008
Balmer: $m=$										
3			.45692	1.6857	.90692	1.0673	.5465	.2157		
4			.61684	1.3800	1.0668	.9578			.3057	.1095
5			.69086	1.2796	1.1409	.9159			.1004	.0419
6			.73107	1.2322	1.1811	.8950			.0474	.0209
7			.75531	1.2057	1.2053	.8830			.0265	.0120
8			.77105	1.1893	1.2210	.8754			.0164	.0076
24			.81674	1.1445	1.2667	.8542			.0448	.0212
25			.81719	1.1441	1.2672	.8540			.0004	.0002
10^3			.82245	1.1392	1.2724	.8516			.0049	.0024

1	2	3	4	5	6	7	8	9	10	11
Paschen: $m=$	$f_o=$ .4500 $\times 10^{15}$	$r_o=$ 1.7029Å	$f_a=$ $\times 10^{15}$	$r_a=$ Å	$f_r=$ (2) +(4) = $\times 10^{15}$	$r_r=$ Å	Range $r_a=$ Å	Range $r_r=$ Å	r_a: $m-(m+1)$	r_r: $m-(m+1)$
4			.15992	3.3942	.60992	1.3904	1.4381	.2448		
5			.23394	2.6339	.68394	1.2882			.7603	.1022
6			.27415	2.3696	.72415	1.2401			.2643	.0481
7			.29839	2.2395	.74839	1.2132			.1301	.0269
8			.31413	2.1640	.76413	1.1964			.0755	.0168
25			.36026	1.9751	.81026	1.1506			.1889	.0458
26			.36066	1.9736	.81066	1.1502			.0015	.0004
10^3			.36553	1.9561	.81553	1.1456			.0175	.0046
Brackett: $m=$										
5			.07402	5.6724	.52402	1.5385	2.8018	.2134		
6			.11423	4.2476	.56423	1.4645			1.4248	.0740
7			.13847	3.7362	.58847	1.4240			.5114	.0405
8			.15421	3.4774	.60421	1.3992			.2588	.0248
30			.20196	2.9051	.65196	1.3300			.5723	.0692
31			.20219	2.9029	.65219	1.3297			.0022	.0003
10^3			.20561	2.8706	.65561	1.3251			.0323	.0046
Pfund: $m=$										
6			.04021	8.5201	.49021	1.6085	4.6548	.1733		
7			.06445	6.2209	.51445	1.5575			2.2992	.0510
8			.08019	5.3776	.53019	1.5266			.8433	.0309
47			.13010	3.8948	.58010	1.4377			1.4828	.0889
48			.13016	3.8936	.58016	1.4376			.0012	.0001
10^3			.13159	3.8653	.58159	1.4352			.0283	.0024

1	2	3	4	5	6	7	8	9	10	11
Lyman: $m=$	$f_o=$ 2.5000 $\times 10^{15}$	$r_o=$ .5429 Å	$f_a=$ $\times 10^{15}$	$r_a=$ Å	$f_r=$ (2) +(4) = $\times 10^{15}$	$r_r=$ Å	Range $r_a=$Å	Range $r_r=$Å	r_a: $m-(m+1)$	r_r: $m-(m+1)$
2			2.4674	.5476	4.9674	.3435	.0955	.0334		
3			2.9243	.4890	5.4243	.3239			.0586	.0196
4			3.0842	.4720	5.5842	.3177			.0170	.0062
5			3.1582	.4646	5.6582	.3149			.0074	.0028
6			3.1984	.4606	5.6984	.3134			.0040	.0015
7			3.2227	.4583	5.7227	.3126			.0023	.0008
8			3.2384	.4568	5.7384	.3120			.0015	.0006
9			3.2492	.4558	5.7492	.3116			.0010	.0004
10			3.2569	.4551	5.7569	.3113			.0007	.0003
18			3.2796	.4530	5.7796	.3105			.0021	.0008
10^3			3.2898	.4521	5.7898	.3101			.0009	.0004
Balmer: $m=$										
3			.45692	1.6857	2.9569	.4854	.5465	.0363		
4			.61684	1.3800	3.1168	.4687			.3057	.0167
5			.69086	1.2796	3.1909	.4614			.1004	.0073
6			.73107	1.2322	3.2311	.4575			.0474	.0039
7			.75531	1.2057	3.2553	.4553			.0265	.0022
8			.77105	1.1893	3.2710	.4538			.0164	.0015
24			.81674	1.1445	3.2818	.4528			.0448	.0010
25			.81719	1.1441	3.2896	.4521			.0004	.0007
10^3			.82245	1.1392	3.3224	.4491			.0049	.0030

1	2	3	4	5	6	7	8	9	10	11
Paschen: $m=$	$f_o=$ 2.5000 $\times 10^{15}$	$r_o=$ .5429 Å	$f_a=$ $\times 10^{15}$	$r_a=$ Å	$f_r=$ (2) +(4) = $\times 10^{15}$	$r_r=$ Å	Range $r_a=$Å	Range $r_r=$Å	r_a: $m-$ $(m+1)$	r_r: $m-$ $(m+1)$
4			.15992	3.3942	2.6599	.5209	1.4381	.0955		
5			.23394	2.6339	2.7339	.5115			.7603	.0094
6			.27415	2.3696	2.7742	.5065			.2643	.0050
7			.29839	2.2395	2.7984	.5036			.1301	.0029
8			.31413	2.1640	2.8141	.5017			.0755	.0019
25			.36026	1.9751	2.8603	.4963			.1889	.0054
26			.36066	1.9736	2.8607	.4962			.0015	.0001
10^3			.36553	1.9561	2.8655	.4958			.0175	.0004
Brackett: $m=$										
5			.07402	5.6724	2.5740	.5324	2.8018	.0174		
6			.11423	4.2476	2.6142	.5270			1.4248	.0054
7			.13847	3.7362	2.6385	.5237			.5114	.0033
8			.15421	3.4774	2.6542	.5216			.2588	.0021
30			.20196	2.9051	2.7020	.5155			.5723	.0061
31			.20219	2.9029	2.7022	.5154			.0022	.0001
10^3			.20561	2.8706	2.7056	.5150			.0323	.0004
Pfund: $m=$										
6			.04021	8.5201	2.5402	.5371	4.6548	.0125		
7			.06445	6.2209	2.5644	.5338			2.2992	.0033
8			.08019	5.3776	2.5802	.5316			.8433	.0022
47			.13010	3.6948	2.6301	.5248			1.4828	.0068
48			.13016	3.8936	2.6302	.5248			.0012	.0000
10^3			.13159	3.8653	2.6316	.5246			.0283	.0002

1	2	3	4	5	6	7	8	9	10	11
Lyman: $m=$	$f_o=$ 9.0000 $\times 10^{15}$	$r_o=$ .2311 Å	$f_a=$ $\times 10^{15}$	$r_a=$ Å	$f_r=$ (2) +(4) = $\times 10^{15}$	$r_r=$ Å	Range $r_a=$ Å	Range $r_r=$ Å	r_a: $m-(m+1)$	r_r: $m-(m+1)$
2			2.4674	.5476	11.467	.1966	.0955	.0088		
3			2.9243	.4890	11.924	.1916			.0586	.0050
4			3.0842	.4720	12.084	.1899			.0170	.0017
5			3.1582	.4646	12.158	.1891			.0074	.0008
6			3.1984	.4606	12.198	.1887			.0040	.0004
7			3.2227	.4583	12.223	.1885			.0023	.0002
8			3.2384	.4568	12.238	.1883			.0015	.0002
9			3.2492	.4558	12.249	.1882			.0010	.0001
10			3.2569	.4551	12.257	.1881			.0007	.0001
18			3.2796	.4530	12.280	.1879			.0021	.0002
10^8			3.2898	.4521	12.290	.1878			.0009	.0001
Balmer: $m=$										
3			.45692	1.6857	9.4569	.2236	.5465	.0056		
4			.61684	1.3800	9.6168	.2211			.3057	.0025
5			.69086	1.2796	9.6909	.2200			.1004	.0011
6			.73107	1.2322	9.7311	.2194			.0474	.0006
7			.75531	1.2057	9.7553	.2190			.0265	.0004
8			.77105	1.1893	9.7710	.2188			.0164	.0002
24			.81674	1.1445	9.8167	.2181			.0448	.0007
25			.81719	1.1441	9.8172	.2181			.0004	.0000
10^8			.82245	1.1392	9.8224	.2180			.0049	.0005

1	2	3	4	5	6	7	8	9	10	11
Paschen: $m=$	$f_o=$ 9.0000 $\times 10^{15}$	$r_o=$ .2311 Å	$f_a=$ $\times 10^{15}$	$r_a=$ Å	$f_r=$ (2) +(4) = $\times 10^{15}$	$r_r=$ Å	Range $r_a=$ Å	Range $r_r=$ Å	r_a: $m-$ $(m+1)$	r_r: $m-$ $(m+1)$
4			.15992	3.3942	9.1599	.2282	1.4381	.0032		
5			.23394	2.6339	9.2339	.2272			.7603	.0010
6			.27415	2.3696	9.2742	.2265			.2643	.0007
7			.29839	2.2395	9.2984	.2262			.1301	.0003
8			.31413	2.1640	9.3141	.2259			.0755	.0003
25			.36026	1.9751	9.3603	.2252			.1889	.0007
26			.36066	1.9736	9.3607	.2251			.0015	.0001
10^3			.36553	1.9561	9.3655	.2251			.0175	.0000
Brackett: $m=$										
5			.07402	5.6724	9.0740	.2299	2.8018	.0022		
6			.11423	4.2476	9.1142	.2292			1.4248	.0007
7			.13847	3.7362	9.1385	.2288			.5114	.0004
8			.15421	3.4774	9.1542	.2285			.2588	.0003
30			.20196	2.9051	9.2020	.2277			.5723	.0008
31			.20219	2.9029	9.2022	.2277			.0022	.0000
10^3			.20561	2.8706	9.2056	.2277			.0323	.0000
Pfund: $m=$										
6			.04021	8.5201	9.0402	.2304	4.6548	.0015		
7			.06445	6.2209	9.0644	.2300			2.2992	.0004
8			.08019	5.3776	9.0812	.2297			.8433	.0003
47			.13010	3.8948	9.1301	.2289			1.4828	.0008
48			.13016	3.8936	9.1302	.2289			.0012	.0000
10^3			.13159	3.8653	9.1316	.2289			.0283	.0000

1	2	3	4	5	6	7	8	9	10	11
Lyman: $m=$	$f_o=$ .01 $\times 10^{15}$	$r_o=$ 21.544 Å	$f_a=$ $\times 10^{15}$	$r_a=$ Å	$f_r=$ (2) +(4) = $\times 10^{15}$	$r_r=$ Å	Range $r_a=$ Å	Range $r_r=$ Å	r_a: $m-(m+1)$	r_r: $m-(m+1)$
2			2.4674	.54766	2.4774	.54618	.09558	.09501		
3			2.9243	.48901	2.9343	.48790			.05865	.05828
4			3.0842	.47196	3.0942	.47094			.01705	.10696
5			3.1582	.46456	3.1682	.46358			.00740	.00736
6			3.1984	.46066	3.2084	.45970			.00390	.00388
7			3.2227	.45834	3.2327	.45739			.00232	.00231
8			3.2384	.45686	3.2484	.45592			.00148	.00147
9			3.2492	.45584	3.2592	.45491			.00102	.00101
10			3.2569	.45512	3.2669	.45420			.00072	.00071
18			3.2796	.45302	3.2896	.45210				
10^3			3.2898	.45208	3.2998	.45117				
Balmer: $m=$										
3			.45692	1.6857	.46692	1.6615	.5465	.5315		
4			.61684	1.3800	.62684	1.3653			.3057	.2962
5			.69086	1.2796	.70086	1.2674			.1004	.0979
6			.73107	1.2322	.74107	1.2211			.0474	.0463
7			.75531	1.2057	.76531	1.1952			.0265	.0259
8			.77105	1.1893	.78105	1.1791			.0164	.0161
15			.80783	1.1529	.81783	1.1435				
16			.80960	1.1512	.81960	1.1418			.0017	.0017
10^3			.82245	1.1392	.83245	1.1300				

1	2	3	4	5	6	7	8	9	10	11
Paschen: $m=$	$f_o=$ .01 $\times 10^{15}$	$r_o=$ 21.544 Å	$f_a=$ $\times 10^{15}$	$r_a=$ Å	$f_r=$ (2) +(4) = $\times 10^{15}$	$r_r=$ Å	Range $r_a=$ Å	Range $r_r=$ Å	r_a: $m-(m+1)$	r_r: $m-(m+1)$
4			.15992	3.3942	.61992	3.2596	1.4381	1.3384		
5			.23394	2.6339	.24394	2.5614			.7603	.6982
6			.27415	2.3696	.28415	2.3137			.2643	.2477
7			.29839	2.2395	.30839	2.1908			.1301	.1229
8			.31413	2.1640	.32413	2.1193			.0755	.0715
25			.36026	1.9751	.37026	1.9394				
26			.36066	1.9736	.37066	1.9380			.0015	.0014
10^3			.36553	1.9561	.37553	1.9212				
Brackett: $m=$										
5			.07402	5.6724	.08402	5.2129	2.8018	2.4318		
6			.11423	4.2476	.12423	4.0165			1.4248	1.1964
7			.13847	3.7362	.14847	3.5665			.5114	.4500
8			.15421	3.4774	.16421	3.3348			.2588	.2317
30			.20196	2.9051	.21196	2.8130				
31			.20219	2.9029	.21219	2.8109			.0022	.0021
10^3			.20561	2.8706	.21561	2.7811				
Pfund: $m=$										
6			.04021	8.5201	.05021	7.3475	4.6548	3.6664		
7			.06445	6.2209	.07445	5.6505			2.2992	1.6970
8			.08019	5.3776	.09019	4.9723			.8433	.6782
47			.13010	3.8948	.14010	3.7072				
48			.13016	3.8936	.14016	3.7061			.0012	.0011
10^3			.13159	3.8653	.14159	3.6811				

12. "Fields"

We have presented evidence which refutes the concept of "action at a distance" in favor of *direct action* (postulates #6 and #7). The former concept requires a series of strange coincidences:

1. If we can presume that the proton and the electron are composed of lesser units—emags and *other emags*—current thought requires that the "charge" of the massive proton be *exactly* equal and opposite to the "charge" of the electron.
2. If we should content ourselves with equal and opposite *surface* "charges" of the proton and electron, ignoring their interiors as being self-neutralizing, we must require that emags or *other emags* align themselves one layer thick on the surfaces of the proton and the electron.

These coincidences tax credibility.

Now there is no reason to believe that all electrons are identical, nor that all protons are identical, in their own dimensions. However, there is no significant error introduced by our assuming their respective identicalness in our dimensions. We may assume, with reasonable confidence, that the structures of electrons, protons, and other subatomic particles (discovered or as yet undiscovered) are constantly participating in collisions of varying magnitudes, and are in a state of *dynamic equilibrium,* within limits.

We speak, glibly, of force "fields," gravitational "fields," electrical "fields," and magnetic "fields." What is a "field"? A "field" is a *zone of influence,* depending upon the source of the force. We have equations quantitating the "field," and these equations *work.* Why, then, are we directing our attention to this subject?

Implicit in the "field" concept is the notion that "fields" are due to forces "acting at a distance." But, according to the emag hypothesis, we have abandoned "action at a distance." Clearly, since the "field" equations do work, we must ask ourselves, "How?" As we have indicated earlier, the idea of "action at a distance" smacks of miracles. We are, therefore, obliged to explain the forces which have hitherto been described as "field" forces in a manner which is compatible with the emag hypothesis.

We have already described, according to the emag hypothesis, the workings of gravitational force "fields" as due to the direct action of the sea of emags. Let us attempt to do the same for electric force "fields." Place two plate electrodes parallel to one another and attach each to the terminal of a battery. A coulomb force "field" is established. This can be demonstrated very easily, and the Millikan experiment is the most famous demonstration. Reversing the "charges" on the plates reverses the direction of the force, and a suspended "charged" particle will have its direction of motion reversed by the above maneuver.

Traditionally, the above experiment is considered proof positive of "action at a distance." But the emag hypothesis requires otherwise. Electric "fields" can exist even in a so-called vacuum. The emag hypothesis states that there is no such thing as a vacuum, except for the time preceding the first *small bang;* that what we consider a vacuum is in reality the sea of emags constituting the ether; and that the "charged" particle suspended between the "charged" plate electrodes is being *pushed* by a rain of *other emags*[1] from one electrode towards the other electrode. Is the push by emags or *other emags?* That all depends upon how we state our *convention.* Let us, therefore, state now, arbitrarily, that what we call the negative electrode rains *other emags.* It could, of course, be the other way around, but as we have said, this is our arbitrary definition.

Returning to the Millikan setup, what happens to the suspended oil droplet with one or more electrons attached? For simplicity, let us eliminate the oil droplet and deal only with the electron. This is similar to the experiment of *time of flight,* in determining the mass of the electron. An electron, which we have considered, heretofore, to be a particle with a magical, negative, electrical charge, was the name adopted for this particle by Stoney, in recognition that the first electrical phenomena–involving static electricity–were demonstrated with the use of amber. The Greek word for amber is electron. In those days, the phenomena truly appeared magical.

Let us take this magical unit electrical "charge," which is suspended between two "charged" plate electrodes, and watch it move away from the so-called cathode, in the direction of the so-called anode. This is a demonstration of the traditional belief that like "charges" repel and unlike "charges" attract. The emag hypothesis suggests that the so-called cathode rains *other emags* (of negative mass) against the electron, which *must also have negative mass* (see table at the end of Chapter 13), and propels the electron towards the so-called anode. The mass of the electron remains fairly constant,

1. We consider an electric current to be a flow of *other emags* (of negative mass), rather than of electrons (of negative mass). See the end of Chapter 19.

even though it is picking up an additional small negative mass from the *other emags* originating at the cathode (see Fig. 13).

In the time of flight experiment, when the "charges" on the electrodes are reversed, the ionized hydrogen which is introduced into the system permits the proton to migrate in the direction previously taken by the electron. In this instance, the proton, which *must have positive mass,* is *propelled* by the rain of *other emags* towards the cathode, losing a small amount of positive mass, momentarily. *The results of the experiment have not been changed. The quantitation of the experiment has not been changed.* But now we can explain electrical phenomena by *direct action,* without resorting to magical "charges" or "action at a distance." Remember, $F = ma$, and since positive mass is reduced and F is constant, a is towards the cathode.

13. Antimatter

The time of flight experiment helps us to determine the relative masses of the proton and electron by virtue of the fact that the work done by the "charged" plate electrodes over the distance between them equals the kinetic energy of the electron or proton between two sensors over another known distance:

$$Fd = \tfrac{1}{2}mv^2$$

The work done by the electrodes, Fd, is the same, whether the projectile is a proton or an electron. However, the velocity of each particle, as measured in the above experiment, is equal to the reciprocal of the square root of its mass. Note, however, that the equation does not specify whether the mass is positive or negative. In the past, it has always been assumed that the masses of both particles were positive, and that their "charges" were opposite, as shown by the opposite direction of deflection between two electrodes. The emag hypothesis demands, since magical electrical "charges" are abandoned, that the *masses* are *opposite* and *there is no charge.*

When Dirac anticipated what was later described as the positron, though his equations indicated the property of negative mass, he was too timid to pursue it, and suggested it only in passing, and abandoned the notion. Consequently, when Anderson discovered the positron, he ascribed a positive "charge" to a positively-massed particle. At that time, of course, there was no *convention* of negative mass to an uncharged electron, as we have stated in the last section. The emag hypothesis describes the positron as an uncharged particle of positive mass. The union of a positron and an electron results in *mutual annihilation,* as presently accepted. The demonstration of gamma "rays," in such an annihilation, simply means that there remain high energy particles, after the collision, of emag or *other emag* aggregates, as in other electromagnetic emanations.

So we see that our hesitancy, when first describing the small bang theory in Chapter 1, in using the term antimatter, need no longer exist, since *other*

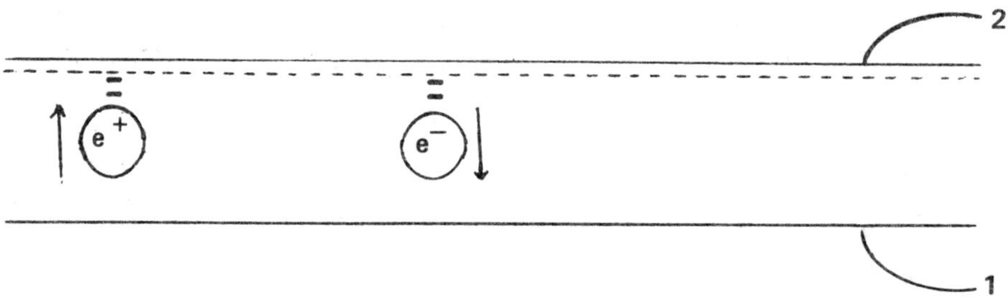

Figure 13 Electrical "field phenomena" by direct action

a. Cathode (2) sends streams of *other emags* to anode (1).
b. Electron, in path of (a), is pushed to anode (1).
c. Electron gains *other emags* from cathode, increasing its negative mass, momentarily.
d. Similarly, a positron in the same electric "field" is accelerated towards the cathode.
e. Both the electron and the positron are aggregates of emags and *other emags*, with the net mass in the former being negative, and in the latter, positive.
f. $F = ma$.
F is the rain of *other emags* from the cathode. Therefore, when the negative-massed electron is struck by the negative force of the rain of *other emags*, the push of the electron is away from the cathode.
When the positive-massed positron is struck by the negative force of the rain of *other emags*, the positive mass is reduced and the direction of the *push* is towards the cathode.

matter is antimatter, when viewed in the proper context. And it is immaterial whether we consider the electron to be negative mass, *other mass,* or anti-mass; or whether we consider the positron to be negative mass, *other mass,* or anti-mass. We have elected the former, arbitrarily, because we feel it would be less disruptive if we switch our thinking of an electron from negative "charge" to negative mass, rather than the other way around. Consequently, our basic particles are

	Mass	Charge
Electron	—	0
Positron	+	0
Proton	+	0
Antiproton	—	0
Neutron	?	0
Antineutron	?	0

14. The Neutron

The neutron, as the previous chart shows, presents a problem. Is it matter or antimatter? If it is not influenced by emag bombardment, as in the manner of electrons or protons, is it composed of mass or negative mass? Evidently, it cannot be either. To be uninfluenced by emag bombardment, it must have no mass! Well, then, what is it? When Bothe and Becker discovered the emanations that were later categorized as neutrons by Chadwick, they assumed these emanations to be gamma "rays." At that time, gamma "rays" (as all electromagnetic emanations) were considered to have no mass. The emag hypothesis demands that *all* emanations are particles and that *all* particles have mass. So even gamma "rays" have mass. Then why are not the presumed gamma "rays" of Bothe and Becker, or the neutrons of Chadwick, deflected in a magnetic "field," as demonstrated in a cloud chamber? Because the radius of curvature is extremely large. In the short distance that a "neutron" travels in a cloud chamber, since it is a very high energy emanation, despite emag and *other emag* bombardment, the radius of curvature is not evident to us. This is also the case with ultraviolet "rays" and "x-rays." As a matter of fact, Einstein's prediction of the bending of light in a large gravitational "field" such as the sun, which was confirmed by observation during a total eclipse of the sun, is evidence of the same phenomenon. We are obliged to conclude, then, that Chadwick's idea of a chargeless mass, though very welcome at the time, can no longer be accepted. What we know as the neutron is better treated, as originally suggested by Bothe and Becker, as a high energy gamma "ray." The fact that protons can be "bumped" out of paraffin (a proton rich donor) by these emanations, when these emanations are ejected from beryllium by alpha particles from polonium, is merely another instance of the Compton effect. It does not require a solid "neutron" to do this, just as it does not require a solid "photon" to eject an electron from a strip of mica foil.

15. The Atomic Nucleus

If we do not have a solid "neutron," what is it that joins with protons, in nuclei, to give the characteristic atomic mass? We have shown that time of flight experiments give values for relative masses of atomic particles. Deuterium has an accepted mass twice that of hydrogen. This greater mass is attributed to the presence of a "neutron" plus a proton in the nucleus. But, if there is no solid particle called a "neutron," the extra mass of the deuterium nucleus must be attributed to an extra proton. "Well, then," we might ask, "why does deuterium only have a single orbital electron?"

This question presupposes that protons and electrons have charges, and that a "neutral" atom must have as many orbital electrons as the nucleus has protons. But we have abandoned charged particles. Therefore, we are not committed to the equality of orbital electrons with nuclear protons. Chance determines how many electrons are pushed into orbit by the emag sea, but not chance alone. Since every orbiting satellite in any planetary system (including the atom) is in some sort of equilibrium state, it is the resultant of all the emag currents in the vicinity which determines the location of the satellite.

The emag sea is all around us, just as our atmosphere is all around the Earth. But just as our atmosphere has currents and eddies, so does the emag sea. Any electron, for example, is subject to emag pushes of various degrees and magnitudes, from all directions. However, when that electron is in the vicinity of a larger mass (the nucleus), it lies in the "shadow" of that larger mass and is sheltered, somewhat, from emag pushes originating from the other side of the larger mass. The size of this umbra, plus the force of the emag currents, determine whether the orbiting electron will stay in orbit (the stationary state of Bohr), or will be pushed into a tighter, higher energy orbit (the Balmer and Lyman series), or be thrust out of orbit because the umbra of the nucleus does not shelter the orbiting electron sufficiently against a very strong "gulf stream" current of the emag sea (ionization).

It must be that a single proton, or two protons in the nucleus, can protect a single orbiting electron, usually. Surely, tritium, with three nuclear protons, can protect a single orbiting electron, but such a nucleus can also protect two orbiting electrons, in which case we no longer have isotopic hydrogen, but isotopic helium, ${}_2He^3$. However, ${}_2He^3$ is unstable, since three nuclear protons can protect two orbiting electrons only with difficulty. Normal helium, ${}_2He^4$, is more stable, etc.

16. Intranuclear Forces

Having abandoned charges, we no longer need to propose huge intranuclear forces to prevent the repulsive destruction of the nucleus by its own protons. Nuclei, if stable, *are* stable because they are not subjected to disruptive forces of any kinds. Nuclei are aggregates of protons (and possibly antiprotons) which happen to have grouped together *by chance. Our whole thesis is based upon haphazard, chance developments that have occurred without a guiding or creative force and which may take tremendous time periods, as reckoned by man.*

If this analysis is correct, it would appear that smaller nuclei would tend to be more stable from disruptive collision with other particles, simply because they offer a smaller target. The reality of the matter is that radioactive nuclei *are* heavy nuclei, offering a large target to disruptive collisions. Implied, of course, is the notion that radioactive disintegration is *not* a spontaneous process.

In dealing with subatomic collisions, calculations of energy changes are based upon the belief that a target nucleus (i.e., beryllium nucleus) struck by a projectile (i.e., an alpha particle from plutonium) can be considered to be at rest (as compared to the projectile's velocity). This is an unjustified assumption. The protons in the beryllium nucleus are in constant motion. Velocity is not in the least related to distance travelled, unless the time interval is specified. Though the distance the beryllium proton travels within its own nucleus may be small, the velocity may be great.

It is also assumed, generally, that interatomic collisions are elastic collisions. This is also an unjustified assumption, however convenient it might be. We have proposed that subatomic particles are collections of intertwining strands of emags and *other emags,* in the manner of tumbleweed. Such a particle would not be expected to participate in an elastic collision. But calculations based on the above assumptions and the relativity factor gamma $(1/(1-v^2/c^2)^{1/2})$ seem to work. This merely suggests that the approximation may be tolerable.

17. "Binding Energy" or Mass Defect

The mass defect of atoms larger than hydrogen, is often described as the "binding energy" of such atoms, and is attributed to the energy holding the nucleus together (intranuclear force). It is presumed that just such an amount of energy needs to be added to the nucleus in order to disrupt it. The mass defect is, of course, the difference in atomic mass as determined by mass spectrography, for example, and the sum of the presumed component parts of that atomic nucleus, i.e., protons, "neutrons," and electrons, the latter sum always being greater. Such a mass defect is a reality, but the above interpretation would appear to be erroneous, in the light of the emag hypothesis.

Initially, the emag hypothesis cannot tolerate disembodied energy, postulate #5. Secondly, we have cast doubt upon the existence of the "neutron" as a solid, subatomic particle. Thirdly, we consider the electron to be of negative mass. Let us, then, interpret the mass defect in terms of the emag hypothesis.

What is the mass of the hydrogen atom? According to our source,[1] ${}_1H^1$ is $1.67339 \pm 0.00031 \times 10^{-24}$ gm. Of course, such a mass must be calculated, not measured. It results from the addition of the mass of the hydrogen nucleus (the proton), which can be measured, and the mass of the electron, which can also be measured in terms of the proton mass, as in the *time of flight* experiment. Such experimental findings tell us that the mass of the electron is 1/1837 times the mass of the proton. The mass of the hydrogen atom, therefore, is believed to be due to the sum of the masses of 1 proton plus 1/1837 proton:

$$ {}_1H^1 = m(\text{proton}) + m(\text{electron}) $$

$$ = m(\text{proton}) + m(1/1837 \text{ proton}) $$

1. *Handbook of Chemistry & Physics*, 49th Edition, p. A-10

$$= 1838/1837\ m(\text{proton})$$

$$= 1.67339 \times 10^{-27}\ \text{Kgm.}$$

The mass of the proton, determined experimentally, is

$$m(\text{proton}) = 1837/1838 \times 1.67339 \times 10^{-27}\ \text{Kgm}$$

$$= 1.67248 \times 10^{-27}\ \text{Kgm, and}$$

$$m(\text{electron}) = -\ 1/1837 \times 1.67248 \times 10^{-27}\ \text{Kgm}$$

$$= -\ 9.10441 \times 10^{-31}\ \text{Kgm}$$

Since, according to the emag hypothesis, the electron has negative mass, the mass of the hydrogen atom should be

$${}_1H^1 = 1.67248 \times 10^{-27}\ \text{Kgm} - 9.10441 \times 10^{-31}\ \text{Kgm}$$

$$= 1.67157 \times 10^{-27}\ \text{Kgm.}$$

This, we will designate 1 AMU.

Dividing the mass of the proton by the mass of 1 AMU, gives us the number of AMUs per proton:

$$m(\text{proton}) = \frac{1.67248 \times 10^{-27}\ \text{Kgm}}{1.67157 \times 10^{-27}\ \text{Kgm}} = 1.00054\ \text{AMU}$$

Similarly,

$$m(\text{electron}) = \frac{-\ 9.10441 \times 10^{-31}\ \text{Kgm}}{1.67157 \times 10^{-27}\ \text{Kgm}} = -\ 5.44662 \times 10^{-4}\ \text{AMU}$$

In the past, the hydrogen atomic mass was called 1 AMU and then the oxygen atomic mass was called 16 AMU, and, finally, the carbon atomic mass of ${}_6C^{12}$ was called 12 AMU, because the atomic masses of the other elements were more nearly whole numbers and were also closer to their corresponding mass numbers. Let us see where our new AMU takes us.

${}_1H^2$ has an atomic number (Z number) of 1 and atomic mass number of 2. The atomic mass is attributed to the sum of 1 proton and 1 neutron in the nucleus. But we no longer have a neutron! How, then, do we get an atomic mass of (approximately) 2, by mass spectrographic analysis? If we realize, according to the emag hypothesis, that we no longer have *attractive* forces, and that, therefore, we no longer need "neutral" atoms, we can get an atomic mass of 2 by using 2 *protons*. Consequently, we shall describe the atom of $_1H^2$ as:

$$_1H^2 = 2 \text{ protons} + 1 \text{ electron}$$

$$= 2 \times 1.67248 \times 10^{-27} \text{ Kgm} - 9.10441 \times 10^{-31} \text{ Kgm}$$

$$= 3.34405 \times 10^{-27} \text{ Kgm.}$$

Dividing this by the mass of 1 AMU,

$$_1H^2 = \frac{3.34405 \times 10^{-27} \text{ Kgm}}{1.67157 \times 10^{-27} \text{ Kgm}}$$

$$= 2.00054 \text{ AMU}$$

Similarly,

$$_2He^3 = 3 \text{ protons} + 2 \text{ electrons}$$

$$= 3 \times 1.00054 \text{ AMU} + 2 \times (-5.44662) \times 10^{-4} \text{ AMU}$$

$$= 3.00053 \text{ AMU}$$

$$_2He^4 = 4 \times 1.00054 + 2 \times (-5.44662) \times 10^{-4}$$

$$= 4.00107 \text{ AMU}$$

$$_6C^{12} = 12 \times 1.00054 + 6 \times (-5.44662) \times 10^{-4}$$

$$= 12.0032 \text{ AMU}$$

$$_8O^{16} = 16 \times 1.00054 + 8 \times (-5.44662) \times 10^{-4}$$

$$= 16.0043 \text{ AMU}$$

$$_{11}Na^{23} = 23 \times 1.00054 + 11 \times (-5.44662) \times 10^{-4}$$

$$= 23.0064 \text{ AMU}$$

$$_{17}Cl^{35} = 35 \times 1.00054 + 17 \times (-5.44662) \times 10^{-4}$$

$$= 35.0096 \text{ AMU}$$

$$_{26}Fe^{56} = 56 \times 1.00054 + 26 \times (-5.44662) \times 10^{-4}$$

$$= 56.0161 \text{ AMU}$$

$$_{88}Ra^{220} = 220 \times 1.00054 + 88 \times (-5.44662) \times 10^{-4}$$

$$= 220.0671 \text{ AMU}$$

$$_{92}U^{238} = 238 \times 1.00054 + 92 \times (-5.44662) \times 10^{-4}$$

$$= 238.078 \text{ AMU}$$

We find that these calculated atomic masses are very close to whole numbers and also very close to the mass numbers. Our figures indicate, however, that our calculated atomic masses are still slightly greater than the observed atomic masses, and, therefore, there still exist small mass defects. Could this be attributed to the experimental and arithmetic inaccuracies? This could surely be the case. There is another consideration. There is no magic reason that the orbital electrons need to match the *Z* number. For one thing, in the absence of charges and neutrons, there is no longer a *Z* number. Furthermore, for "neatness" there could very well be as many orbital electrons in an atom as there are protons in the nucleus. This is surely the case in $_1H^1$, and there is no reason that it could not be the case in the other atoms. If such were the case, and of course this is strictly conjecture, the atomic masses would be perfect whole numbers.

But let us continue to assume that there still remains a small mass defect. Disrupted nuclei are sources of beta emanations, gamma emanations, neutrinos, and antineutrinos. Some such emanations are presently presumed to be massless. However, the emag hypothesis demands that all such emanations of high energy do have mass. If we consider, too, antimesons and antielectrons (positrons), all such masses, positive and negative in total, can, evidently, be the equivalent of any mass defect. Column #9, in Table 3B, indicates the number of electron masses in any particular mass defect. The implied conclusion is clear.

Many books display a curve of mass defect per baryon as a smooth curve, which, after the initial spike of $_2He^4$ and the subsequent dip of $_2He^5$, reaches a smooth maximum at $_{26}Fe^{56}$. It is not inappropriate, in presenting an empirical curve, to smooth it out. However, if we do not smooth it out, we notice various saw-toothed jogs. A glance at Table 3A indicates the absence of a smooth progression of mass defect with increasing mass number. Examples of this are seen in $_2He^4$, in column #9, and in column #10 for $_2He^5$, $_3Li^6$, $_7N^{14}$, and $_9F^{19}$ Table 3A represents current thought, where 1 AMU is 1/12 the mass of $_6C^{12}$, and where column #8 is the sum of positive masses (neutron, proton, and electron) (see Fig. 14).

Table 3B represents emag hypothesis thinking, where $_1H^1$ is 1 AMU, and is composed of the sum of 1 positive mass proton and 1 negative mass

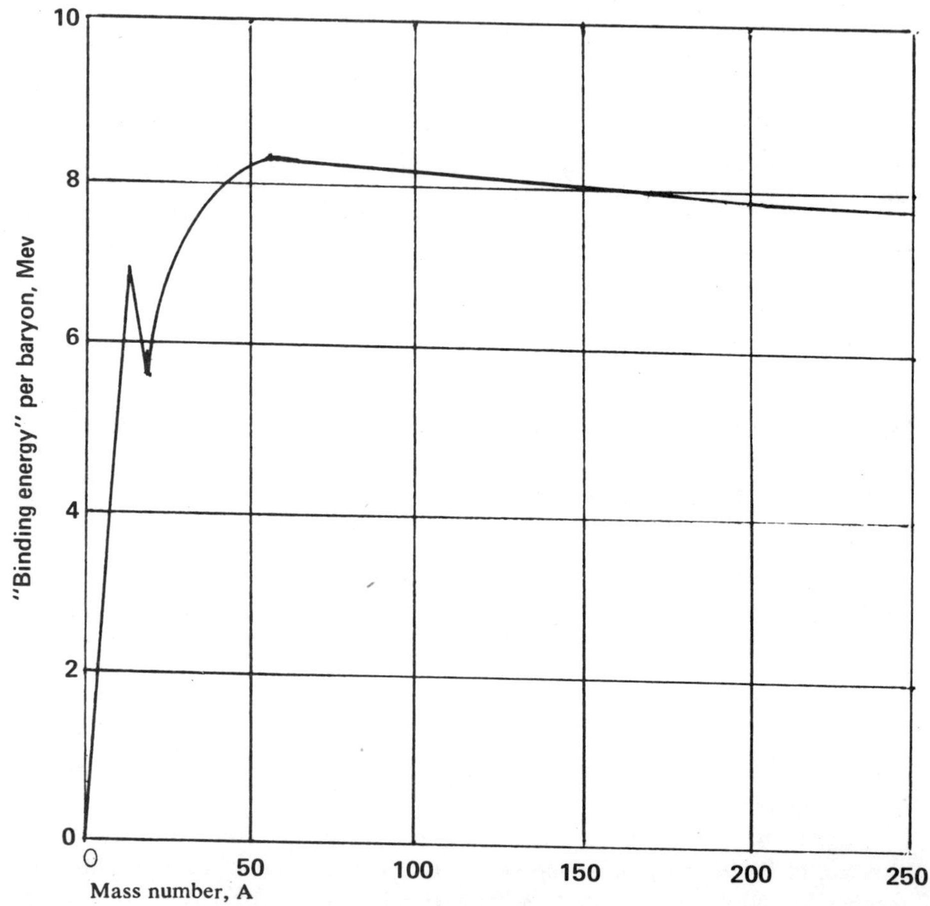

Figure 14. "Binding energy" per baryon

Table 3A

1	2	3	4	5	6	7	8	9	10
#	Z#	Element	*N* 1.00865 AMU	*P* 1.007277 AMU	*E* 5.48327×10^{-4} AMU	Handbook mass if ${}_{6}C^{12}$ is 12.00000AMU	Calculated mass by parts in AMUs	Mass defect in AMUs	Mass defect per baryon in AMUs
1	1	${}_{1}H^{1}$	0	1	1	1.007825	1.007825	0	0
2	1	${}_{1}H^{2}$	1	1	1	2.01410	2.01648	.00238	.00119
3	2	${}_{2}He^{3}$	1	2	2	3.01603	3.02430	.00827	.00276
4	2	${}_{2}He^{4}$	2	2	2	4.00260	4.03295	.03035	.00760
5	2	${}_{2}He^{5}$	3	2	2	5.01230	5.04160	.02930	.00586
6	3	${}_{3}Li^{6}$	3	3	3	6.01513	6.04942	.03430	.00572
7	3	${}_{3}Li^{7}$	4	3	3	7.01601	7.05808	.04206	.00601
8	4	${}_{4}Be^{9}$	5	4	4	9.01219	9.07455	.06236	.00693
9	5	${}_{5}B^{11}$	6	5	5	11.0093	11.0910	.0817	.00743
10	6	${}_{6}C^{12}$	6	6	6	12.0000	12.0988	.0988	.00824
11	7	${}_{7}N^{14}$	7	7	7	14.0031	14.1153	.1122	.00802
12	8	${}_{8}O^{16}$	8	8	8	15.9949	16.1318	.1369	.00856
13	9	${}_{9}F^{19}$	10	9	9	18.9984	19.1569	.1585	.00834
14	10	${}_{10}Ne^{20}$	10	10	10	19.9924	20.1648	.1724	.00862
15	11	${}_{11}Na^{23}$	12	11	11	22.9898	23.1899	.2001	.00870
16	13	${}_{13}Al^{27}$	14	13	13	26.9815	27.2228	.2413	.00894
17	15	${}_{15}F^{31}$	16	15	15	30.9738	31.2558	.2820	.00910
18	17	${}_{17}Cl^{35}$	18	17	17	34.9688	35.2887	.3199	.00914
19	19	${}_{19}K^{39}$	20	19	19	38.9637	39.3219	.3580	.00918
20	21	${}_{21}Si^{45}$	24	21	21	44.9559	45.3719	.4160	.00924
21	23	${}_{23}V^{51}$	28	23	23	50.9441	51.4222	.4781	.00937
22	25	${}_{25}Mn^{55}$	30	25	25	54.9381	55.4551	.5170	.00940
23	26	${}_{26}Fe^{56}$	30	26	26	55.9349	56.4630	.5280	.00943
24	28	${}_{28}Ni^{58}$	30	28	28	57.9353	58.4786	.5433	.00937
25	30	${}_{30}Zn^{64}$	34	30	30	63.9291	64.5288	.5998	.00937
26	33	${}_{33}As^{75}$	42	33	33	74.9216	75.6215	.6999	.00933
27	36	${}_{36}Kr^{84}$	48	36	36	83.9115	84.6969	.7854	.00935
28	39	${}_{39}Y^{89}$	50	39	39	88.9054	89.7377	.8323	.00935
29	42	${}_{42}Mo^{98}$	56	42	42	97.9055	98.8131	.9076	.00926
30	45	${}_{45}Rh^{103}$	58	45	45	102.905	103.854	.9488	.00921
31	48	${}_{48}Cd^{114}$	66	48	48	113.904	114.946	1.042	.00914
32	51	${}_{51}Sb^{121}$	70	51	51	120.904	122.004	1.100	.00910
33	54	${}_{54}Xe^{132}$	78	54	54	131.904	133.097	1.193	.00904
34	57	${}_{57}La^{139}$	82	57	57	138.906	140.155	1.249	.00899
35	60	${}_{60}Nd^{142}$	82	60	60	141.908	143.179	1.271	.00895
36	63	${}_{63}Eu^{153}$	90	63	63	152.921	154.271	1.350	.00883
37	66	${}_{66}Dy^{164}$	98	66	66	163.929	165.364	1.435	.00875
38	69	${}_{69}Tm^{169}$	100	69	69	168.934	170.405	1.471	.00870
39	72	${}_{72}Hf^{180}$	108	72	72	179.947	181.498	1.551	.00861
40	75	${}_{75}Re^{187}$	112	75	75	186.956	188.556	1.600	.00855
41	78	${}_{78}Pt^{195}$	117	78	78	194.965	196.622	1.657	.00850
42	81	${}_{81}Tl^{205}$	124	81	81	204.974	206.706	1.732	.00845

Table 3A (contd.)

1	2	3	4	5	6	7	8	9	10
#	Z#	Element	*N* 1.00865 AMU	*P* 1.007277 AMU	*E* 5.48327×10^{-4} AMU	Handbook mass if $_6C^{12}$ is 12.00000AMU	Calculated mass by parts in AMUs	Mass defect in AMUs	Mass defect per Baryon in AMUs
43	83	$_{83}Bi^{209}$	126	83	83	208.980	210.739	1.759	.00842
44	88	$_{88}Ra^{220}$	132	88	88	220.010	221.830	1.820	.00827
45	90	$_{90}Th^{223}$	133	90	90	223.021	224.854	1.834	.00822
46	92	$_{92}U^{238}$	146	92	92	238.051	239.983	1.932	.00812
45	90	$_{90}Th^{223}$	133	90	90	223.021	224.855	1.834	.00822
47	90	$_{90}Th^{224}$	134	90	90	224.021	225.863	1.842	.00822
48	90	$_{90}Th^{225}$	135	90	90	225.024	226.872	1.848	.00821
49	90	$_{90}Th^{226}$	136	90	90	226.025	227.881	1.856	.00821
50	90	$_{90}Th^{227}$	137	90	90	227.028	228.889	1.861	.00820
51	90	$_{90}Th^{228}$	138	90	90	228.029	229.899	1.869	.00820
52	90	$_{90}Th^{229}$	139	90	90	229.032	230.907	1.875	.00819
53	90	$_{90}Th^{230}$	140	90	90	230.033	231.915	1.882	.00818
54	90	$_{90}Th^{232}$	142	90	90	232.038	233.932	1.894	.00817
55	92	$_{92}U^{227}$	135	92	92	227.031	228.888	1.857	.00818
56	92	$_{92}U^{228}$	136	92	92	228.031	229.896	1.865	.00818
57	92	$_{92}U^{229}$	137	92	92	229.033	230.905	1.872	.00817
58	92	$_{92}U^{230}$	138	92	92	230.034	231.914	1.880	.00817
59	92	$_{92}U^{232}$	140	92	92	232.037	233.931	1.894	.00816
60	92	$_{92}U^{233}$	141	92	92	233.040	234.940	1.900	.00815
61	92	$_{92}U^{234}$	142	92	92	234.041	235.948	1·907	.00815
62	92	$_{92}U^{235}$	143	92	92	235.044	236.957	1.913	.00814
63	92	$_{92}U^{236}$	144	92	92	236.046	237.966	1.920	.00813
64	92	$_{92}U^{238}$	146	92	92	238.051	239.983	1.932	.00812
65	93	$_{93}Np^{231}$	138	93	93	231.038	232.921	1.883	.00815
66	93	$_{93}Np^{233}$	140	93	93	233.041	234.939	1.898	.00814
67	93	$_{93}Np^{236}$	143	93	93	236.047	237.965	1.918	.00812
68	93	$_{93}Np^{237}$	144	93	93	237.048	238.973	1.925	.00812

NOTES—Table 3A

Column 1: Identification #'s of lines
Column 2: Z number
Column 3: Element and/or isotope
Column 4: number of "neutrons" in nucleus, of mass 1.00865 AMU each
Column 5: number of protons in nucleus (current belief) of mass 1.007277 AMU each
Column 6: number of electrons around nucleus, of mass 5.48327×10^{-4} AMU each
Column 7: mass of elements in column 3, if $_6C^{12}$ is 12.00000 AMU
Column 8: calculated mass of elements in column 3: the sum of the parts in columns 4, 5, and 6
Column 9: mass defect: the difference between column 8 and column 7
Column 10: mass defect per baryon: column 9 divided by the sum of columns 4 and 5

Table 3B

1 \#	2 Z#	3 Element	4 P 1.00054 AMU	5 −E 5.44662 × 10⁻⁴ AMU	6 Calculated mass from Handbook if $_{1}H^{1}$ is 1 AMU	7 Calculated mass by parts in AMU's	8 Mass defect in AMU's	9 \# E's	10 Mass defect per baryon in AMU's
1	1	$_{1}H^{1}$	1	1	1.00000	1.00000	0	0	0
2	1	$_{1}H^{2}$	2	1	1.99846	2.00054	.00208	3.819	.00104
3	2	$_{2}He^{3}$	3	2	2.99261	3.00053	.00792	14.54	.00264
4	2	$_{2}He^{4}$	4	2	3.97152	4.00107	.02955	54.25	.00739
5	2	$_{2}He^{5}$	5	2	4.97338	5.00161	.02823	51.83	.00565
6	3	$_{3}Li^{6}$	6	3	5.96843	6.00161	.03318	60.92	.00553
7	3	$_{3}Li^{7}$	7	3	6.96154	7.00215	.04061	74.56	.00580
8	4	$_{4}Be^{9}$	9	4	8.94222	9.00268	.06046	111.0	.00672
9	5	$_{5}B^{11}$	11	5	10.9238	11.0032	.0794	145.8	.00722
10	6	$_{6}C^{12}$	12	6	11.9068	12.0032	.0964	177.0	.00803
11	7	$_{7}N^{14}$	14	7	13.8943	14.0037	.1094	200.8	.00781
12	8	$_{8}O^{16}$	16	8	15.8707	16.0043	.1335	245.1	.00834
13	9	$_{9}F^{19}$	19	9	18.8509	19.0054	.1545	283.7	.00813
14	10	$_{10}Ne^{20}$	20	10	19.8372	20.0054	.1682	308.8	.00841
15	11	$_{11}Na^{23}$	23	11	22.8113	23.0064	.1951	358.2	.00848
16	13	$_{13}Al^{27}$	27	13	26.7720	27.0075	.2355	432.4	.00872
17	15	$_{15}P^{31}$	31	15	30.7332	31.0086	.2754	505.6	.00888
18	17	$_{17}Cl^{35}$	35	17	34.6973	35.0096	.3123	573.4	.00892
19	19	$_{19}K^{39}$	39	19	38.6612	39.0107	.3495	641.7	.00896
20	21	$_{21}Si^{45}$	45	21	44.6069	45.0129	.4060	745.4	.00902
21	23	$_{23}V^{51}$	51	23	50.5484	51.0150	.4666	856.7	.00915
22	25	$_{25}Mn^{55}$	55	25	54.5115	55.0161	.5046	926.4	.00917
23	26	$_{26}Fe^{56}$	56	26	55.5006	56.0161	.5155	946.4	.00920
24	28	$_{28}Ni^{58}$	58	28	57.4855	58.0161	.5306	974.2	.00915
25	30	$_{30}Zn^{64}$	64	30	63.4327	64.0182	.5855	1075	.00915
26	33	$_{33}As^{75}$	75	33	74.3399	75.0225	.6826	1253	.00910
27	36	$_{36}Kr^{84}$	84	36	83.2600	84.0258	.7658	1406	.00912
28	39	$_{39}Y^{89}$	89	39	88.2151	89.0268	.8117	1490	.00912
29	42	$_{42}Mo^{98}$	98	42	97.1435	98.0300	.8847	1624	.00903
30	45	$_{45}Rh^{103}$	103	45	102.106	103.031	.925	1698	.00898
31	48	$_{48}Cd^{114}$	114	48	113.020	114.035	1.015	1864	.00890
32	51	$_{51}Sb^{121}$	121	51	119.965	121.038	1.073	1970	.00887
33	54	$_{54}Xe^{132}$	132	54	130.880	132.042	1.162	2133	.00880
34	57	$_{57}La^{139}$	139	57	137.827	139.044	1.217	2234	.00876
35	60	$_{60}Nd^{142}$	142	60	140.806	142.044	1.238	2273	.00872
36	63	$_{63}Eu^{153}$	153	63	151.734	153.048	1.314	2412	.00859
37	66	$_{66}Dy^{164}$	164	66	162.656	164.053	1.397	2565	.00852
38	69	$_{69}Tm^{169}$	169	69	167.622	169.054	1.432	2629	.00847
39	72	$_{72}Hf^{180}$	180	72	178.550	180.058	1.508	2769	.00838
40	75	$_{75}Re^{187}$	187	75	185.504	187.060	1.556	2857	.00832
41	78	$_{78}Pt^{195}$	195	78	193.451	195.063	1.612	2960	.00827
42	81	$_{81}Tl^{205}$	205	81	203.382	205.066	1.684	3092	.00821

Table 3B (contd.)

1	2	3	4	5	6	7	8	9	10
#	Z#	Element	P 1.00054 AMU	−E 5.44662×10^{-4} AMU	Calculated mass from Handbook if $_1H^1$ is 1AMU	Calculated mass by parts in AMU's	Mass defect in AMU's	# E's	Mass defect per baryon in AMU's
43	83	$_{83}Bi^{209}$	209	83	207.357	209.068	1.711	3141	.00819
44	88	$_{88}Ra^{220}$	220	88	218.302	220.071	1.769	3248	.00804
45	90	$_{90}Th^{223}$	223	90	221.289	223.071	1.782	3272	.00799
46	92	$_{92}U^{238}$	238	92	236.203	238.078	1.875	3442	.00788
45	90	$_{90}Th^{223}$	223	90	221.289	223.071	1.782	3272	.00799
47	90	$_{90}Th^{224}$	224	90	224.021	224.072	.0505	92.72	.000225
48	90	$_{90}Th^{225}$	225	90	225.024	225.072	.0488	89.60	.000217
49	90	$_{90}Th^{226}$	226	90	226.025	226.073	.0481	88.31	.000213
50	90	$_{90}Th^{227}$	227	90	227.028	227.074	.0458	84.09	.000202
51	90	$_{90}Th^{228}$	228	90	228.029	228.074	.0454	83.35	.000199
52	90	$_{90}Th^{229}$	229	90	229.032	229.075	.0430	78.95	.000188
53	90	$_{90}Th^{230}$	230	90	230.033	230.075	.0421	77.30	.000183
54	90	$_{90}Th^{232}$	232	90	232.038	232.076	.0381	69.95	.000164
55	92	$_{92}U^{227}$	227	92	225.268	227.072	1.804	3312	.00795
56	92	$_{92}U^{228}$	228	92	226.260	228.073	1.813	3329	.00795
57	92	$_{92}U^{229}$	229	92	227.255	229.074	1.818	3339	.00794
58	92	$_{92}U^{230}$	230	92	228.248	230.074	1.826	3352	.00794
59	92	$_{92}U^{232}$	232	92	230.235	232.075	1.840	3378	.00793
60	92	$_{92}U^{233}$	233	92	231.231	233.077	1.845	3387	.00792
61	92	$_{92}U^{234}$	234	92	232.224	234.076	1·852	3400	.00792
62	92	$_{92}U^{235}$	235	92	233.219	235.077	1.858	3411	.00790
63	92	$_{92}U^{236}$	236	92	234.213	236.077	1.864	3422	.00790
64	92	$_{92}U^{238}$	238	92	236.203	238.078	1.875	3442	.00788
65	93	$_{93}Np^{231}$	231	93	229.244	231.074	1.830	3360	.00792
66	93	$_{93}Np^{233}$	233	93	231.232	232.075	1.843	3384	.00791
67	93	$_{93}Np^{236}$	236	93	234.214	236.077	1.863	3420	.00789
68	93	$_{93}Np^{237}$	237	93	235.208	237.077	1.869	3431	.00789

NOTES—Table 3B

Column 4: number of protons in nucleus, baryons, assuming the nonexistence of neutrons, of mass 1.00054 AMU each

Column 5: number of electrons around nucleus, of mass 5.44662×10^{-4} negative AMU each

Column 6: mass of elements in column 3, using $_1H^1$ as 1 AMU, calculated from column 7 of Table A, divided by 1.007825, the mass of $_1H^1$ in Table A

Column 7: calculated mass of elements in column 3: the sum of the protons in column 4 plus the sum of the negative mass of the electrons in column 5

NOTES—Table 3B (contd.)

Column 8: mass defect: the difference between column 7 and column 6
Column 9: the number of electrons the mass defect in column 8 equals
Column 10: mass defect per baryon: column 8 divided by column 4

The mass of the proton in column 4 of Table 3B is determined as follows:

The mass of $_1H^1$ is 1.67339×10^{-27} Kgm, as determined experimentally by adding the mass of a proton to the positive mass of an electron. The mass of the proton can be determined by mass spectrographic analysis. The mass of the electron can be determined, by time of flight experiments, to be 1/1837 the mass of the proton. Since the hydrogen atom is believed to be composed of a proton nucleus and a single orbital electron, the total mass of the hydrogen atom is:

$$m_p = (1837/1838) \times 1.67339 \times 10^{-27} \text{ Kgm}$$
$$= 1.67248 \times 10^{-27} \text{ Kgm}$$

$$m_e = (1/1838) \times 1.67339 \times 10^{-27} \text{ Kgm}$$
$$= -\,9.10441 \times 10^{-31} \text{ Kgm.}$$

However, the real mass of the hydrogen atom is the sum of the positive mass of the proton and the negative mass of the electron:

$$m({}_1H^1) = 1.67248 \times 10^{-27} \text{ Kgm} - 9.10441 \times 10^{-31} \text{ Kgm}$$
$$= 1.67157 \times 10^{-27} \text{ Kgm}$$

which is less than the mass of the proton.

Now we have defined the mass of the hydrogen atom as 1 AMU. Therefore, dividing the mass of the proton by the mass of the hydrogen atom, gives us the mass of the proton in AMU:

$$m_p = \frac{\text{mass of proton}}{\text{mass } {}_1H^1}$$

$$= \frac{1.67248 \times 10^{-27} \text{ Kgm}}{1.67157 \times 10^{-27} \text{ Kgm}}$$

$$= 1.00054 \text{ AMU}$$

Similarly,

$$m_e = \frac{\text{mass of electron}}{\text{mass } {}_1H^1}$$

$$= \frac{-\,9.10441 \times 10^{-31} \text{ Kgm}}{1.67157 \times 10^{-27} \text{ Kgm}}$$

$$= -\,5.44662 \times 10^{-4} \text{ AMU}$$

electron. This gives smoother curves in column #8, which represents mass defect in AMUs, and in column #9, which represents equivalent electron masses. In Table 3B, column #8, we notice a jog for $_2He^5$. If we look to Table 4, we note that $_2He^5$ is radioactive. Also, in Table 4, we see a reduced mass defect, a jog, for $_3Li^5$, which is also radioactive. Evaluating the mass defects of the radioactive isotopes in Table 4 against their neighbors of close atomic mass in Table 3B, we discover that whenever there is a jog in the curve of mass defect, we are dealing with a radioactive isotope. A very dramatic example of this is seen in Table 3B, in the thorium series. The mass defects of the thorium isotopes are relatively miniscule compared to the mass defects of the neighboring elements. In other words, radioactive isotopes have lesser mass defects than expected from their mass numbers, signifying a greater ease of disruption by collisions with other particles. This ease of disruption is the very characteristic which constitutes radioactivity. We observe, in our tables, that mass defect, aside from the jogs, is directly proportional to the mass number, while the mass defect per baryon reaches a second maximum at $_{26}Fe^{56}$. This latter observation implies a maximal stability of iron against all other elements. It must have some greater significance that cobalt, iron, and nickel, all ferromagnetic, are neighbors at the maximum mass defect per baryon.

Let us return to our discussion of radioactivity. It would be nice to find that the mass defects of radioactive isotopes were directly proportional to their half-lives, indicating that a very unstable isotope with a very short half-life would also have a very small mass defect. While this is so in many instances, alas, there are many instances in which this is not the case. We must, therefore, consider the possibility that the atomic masses of short-lived, unstable, radioactive isotopes, as reported, may be erroneously small, and that the atomic masses of long-lived, more stable, radioactive isotopes, as reported, may be erroneously large.

It is obligatory to point out, as we did in the chapter on "fields," that all electrons and all protons are not identical. Since the process of aggregation of emags and *other emags,* matter and antimatter, into filaments, and the filaments into intertwining masses of "tumbleweed," is purely haphazard, and since the numbers of emags and filaments are so great in any proton or electron, the final particle can very well have a range of sizes and masses which our instruments cannot detect. Furthermore, our experimental determinations are relatively few, and averaged. Consequently, our brash computations of *exact* mass defect, is really only an approximation. Also, it is very likely that a typical proton can lose an electron mass and remain a typical proton, or lose one or more or gain one or more electrons or positrons and remain a typical proton.

Table 4

1	2	3	4	5	6	7	8	9	10	11
Z#	Element	Handbook mass if $_6C^{12}$ is 12.00000AMU	Mass defect if $_6C^{12}$ is 12.00000	Calculated mass from Handbook if $_1H^1$ is 1AMU	Mass defect if $_1H^1$ is 1 AMU	*N* 1.00865 AMU	*P* 1.007277 AMU	*E* 5.48327×10^{-4} AMU	*P* 1.00054 AMU	−E 5.44662×10^{-4} AMU
2	$_2He^3$	3.01603	.00827	2.99261	.00792	1	2	2	3	2
2	$_2He^4$	4.00260	.03035	3.97152	.02955	2	2	2	4	2
2	$_2He^5$*	5.01230	.02930	4.97338	.02823	3	2	2	5	2
3	$_3Li^5$*	5.0125	.02828	4.97358	.02749	2	3	3	5	3
3	$_3Li^6$	6.01513	.03430	5.96843	.03318	3	3	3	6	3
4	$_4Be^6$*	6.0198	.02880	5.97306	.02800	2	4	4	6	4
4	$_4Be^8$*	8.00531	.06059	7.94315	.05899	4	4	4	8	4
5	$_5B^9$*	9.01333	.06040	8.94335	.05879	4	5	5	9	5
9	$_9F^{16}$*	16.01171	.11927	15.8874	.11634	7	9	9	16	9
20	$_{20}Ca^{48}$*	47.9524	.44631	47.5797	.43533	28	20	20	48	20
23	$_{23}V^{50}$*	49.9472	.46633	49.5592	.45527	27	23	23	50	23
49	$_{49}In^{115}$*	114.904	1.0503	114.012	1.0234	66	49	49	115	49
57	$_{57}La^{138}$*	137.907	1.2397	136.836	1.2075	81	57	57	138	57
58	$_{58}Ce^{142}$*	141.909	1.2715	140.807	1.2381	84	58	58	142	58
60	$_{60}Nd^{144}$*	143.910	1.2861	142.793	1.2521	84	60	60	144	60
62	$_{62}Sm^{146}$*	146.913	.29877	145.772	.27307	84	62	62	146	62
64	$_{64}Gd^{148}$*	147.918	1.3094	146.770	1.2751	84	64	64	148	64
64	$_{64}Gd^{149}$*	148.919	1.3171	147.763	1.2826	85	64	64	149	64
64	$_{64}Gd^{150}$*	149.918	1.3267	148.754	1.2921	86	64	64	150	64
64	$_{64}Gd^{152}$*	151.920	1.3420	150.740	1.3072	88	64	64	152	64
65	$_{65}Tb^{151}$*	150.923	1.3295	149.751	1.2951	86	65	65	151	65
66	$_{66}Dy^{152}$*	151.924	1.3364	150.744	1.3021	86	66	66	152	66
66	$_{66}Dy^{153}$*	152.925	1.3440	151.738	1.3087	87	66	66	153	66
66	$_{66}Dy^{154}$*	153.925	1.3527	152.730	1.3172	88	66	66	154	66
72	$_{72}Hf^{174}$*	173.940	1.5057	172.589	1.4657	102	72	72	174	72
75	$_{75}Re^{187}$*	186.956	1.5997	185.504	1.5561	112	75	75	187	75
78	$_{78}Pt^{192}$*	191.961	1.6355	190.470	1.5912	114	78	78	192	78
82	$_{82}Pb^{204}$*	203.973	1.7240	202.389	1.6765	122	82	82	204	82
84	$_{84}Po^{206}$*	205.980	1.733	204.381	1.6845	122	84	84	206	84
84	$_{84}Po^{207}$*	206.982	1.7393	205.375	1.6910	123	84	84	207	84
84	$_{84}Po^{208}$*	207.981	1.7544	206.366	1.7006	124	84	84	208	84
84	$_{84}Po^{209}$*	208.982	1.7566	207.359	1.7081	125	84	84	209	84

* radioactive

18. *Chemical Reactions*

It is clear that the Z number can represent the number of orbital electrons, even if we no longer have neutrons. It is equally clear that the periods and groups in the periodic table suggest a number of orbital electrons equal to the Z number; chemical "likeness" being determined, as presently interpreted, by the number of electrons in an unfilled orbit. In the past, it was very "neat" to have the Z number equal to either the number of protons in the nucleus or the number of orbital electrons. In the last section, we suggested another "neat" solution. However, while neatness is desirable, it is no panacea. There is no reason to believe that there is any *special* number of electrons required, though it still seems desirable to have chemical activity related to the number of electrons, from the standpoint of likeness. Surely, chemical likeness is not related to the number of baryons in the nucleus, as isotopes attest. But, as we have indicated in the previous section, the number of electrons orbiting a nucleus is a matter of chance, and the umbra of the nucleus determines, in the last analysis, which isotopes are more stable than others.

How do chemical reactions take place? It is stated that atoms may become ions by gaining or losing one or more electrons. It is stated that nonpolar bonds occur by sharing electrons. Various schemes have been devised to show these actions. Such schemes, however, make the assumption that neutral sodium, for example, can part quite readily with an electron, which may be readily picked up by a nearby chlorine atom, for example. When one stops to think that such "neutral" atoms are not stationary, but racing hither and yon, as do all particles, it would be quite a trick for a racing chlorine atom to "happen" to gain an even faster moving electron which has been forcibly knocked out of sodium orbit. And so, according to present thought, we have ions. No provision seems to have been made for any free electrons which have been knocked out of orbit, and have remained uncaptured orphans.

As far as a nonpolar bond is concerned, it would appear to be an even better trick for two atoms to share one, two, four, or even six electrons, and to continue to have electrons orbiting their own or other nucleus. The above notions beg for improvement. Can the emag hypothesis offer that improvement?

Let us return to "neutral" sodium and "neutral" chlorine atoms. The sodium atom with 23 protons has, let us say, 11 electrons in orbit. Along comes a chlorine atom with 35 protons and, let us say, 17 electrons in orbit. We remember the analysis of the hydrogen spectrum and energy states, where the inner orbits are in higher energy states than the outer orbits. Just as in the case of the single electron of hydrogen, each of the orbital electrons in both sodium and chlorine varies its orbit, within limits, and also its energy state, these orbits and energy states being dependent upon the ambient emag stream at any particular place and time. When the push of the emag sea is sufficiently great, either because of a greater "temperature" or a miniscule violent "explosion" (what we call reaction heats and energies), one or more outer orbital electrons from one atom, perhaps the sodium atom, breaches the outer orbital ring of the other atom, perhaps the chlorine atom, and a loop chain effect is achieved. Now the two orbits are intertwined, and the two atoms travel together as a molecule, yet the electrons of each atom are orbiting their own, respective nuclei. In the same fashion, two or more atoms can form chain "ions." Their intertwined loops may or may not be cross-looped, since each atom may or may not be in the same ambient emag sea, though in our dimensions they are close neighbors, because in their own dimensions, they are huge distances apart. More will be said of this concept, shortly (see Fig. 15).

But we did say "ions," and ions are defined as atoms or groups of atoms, with electric charges, which can migrate in an electric "field." The emag hypothesis says there are no charges. What, then, are ions?

Let us return to the time of flight experiment with hydrogen. With a spark as an energy source, ${}_1H^1$ is "ionized," resulting in a free proton and a free electron, both uncharged, but of opposite mass. When either of these particles is allowed to drift between two "charged" plates of an electric "field," *in vacuo,* the particle is accelerated greatly by *other emag* pressure. If the particle is an electron, it is pushed by the rain of *other emags* from the cathode, gaining negative mass, momentarily. If the particle is a proton, then the rain of *other emags* reduces the positive mass of the proton, momentarily, and since the force is unchanged, there is negative acceleration *towards* the cathode.

Thus we see that *ions are uncharged atoms or groups of atoms which can migrate, perceptibly, in the sea of other emags.* Such ions can migrate, perceptibly, whether the electric "field" is caused by two plate electrodes or by two electrodes in an electrolyte.

We know that various elements have more than one valence. This characteristic is easily explained by the emag hypothesis. The different valences are determined by the number of electron orbits of one atom breached by a number of electrons in orbit around another nucleus. In the case of ferrous iron, for example, it is possible for an electron from each of the outer orbits

Figure 15. Spatial relationships of three chemical bonds

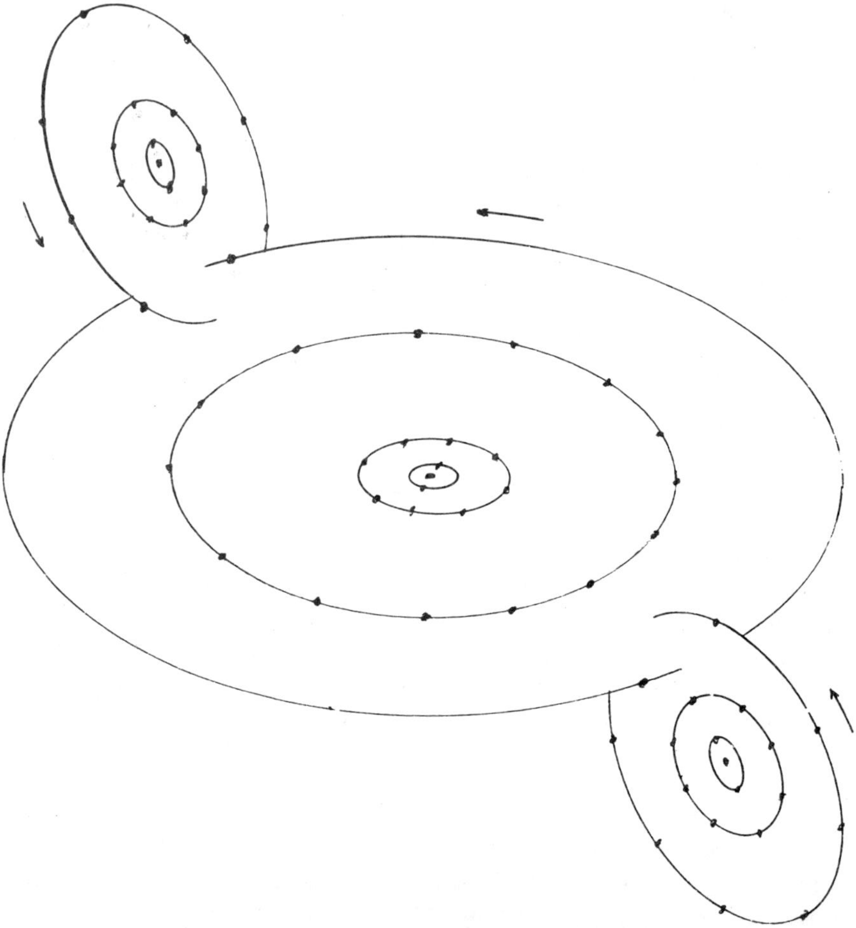

A. Nonscale diagram of $FeCl_2$
Z number Fe 26
Z number Cl 17

The diagrams show possible spatial relationships of the various chemical bonds involved. Only the diagram of HBr is drawn to scale. The directions of the orbital electrons are arbitrary. It should also be pointed out that though the diagrams are drawn with some kind of symmetry, this is not necessary, and there is no reason to believe that the orbital electrons are spaced regularly in their orbits, nor that the chemical bonding

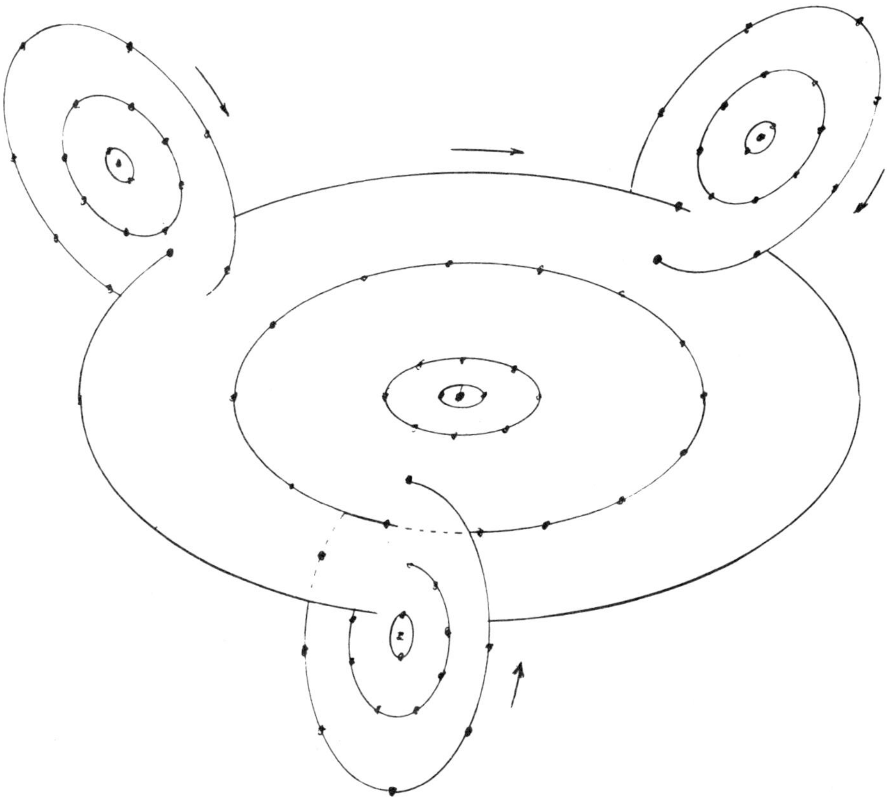

B. Nonscale diagram of $FeCl_3$
Z number Fe 26
Z number Cl 17

is symmetrically arranged, in reality. The question of spatial isomerism is not compromised.

Quantum theory describes the paths of orbital electrons as compatible with the concept that electrons may be considered "clouds" of negative electric charge around a nucleus. Since quantum theory really describes wave function probabilities, and since the frequency of an electron around a nucleus is of the order of 10^{15}/sec., the cloud concept really means that at any particular miniscule part of a second, the electron is at some

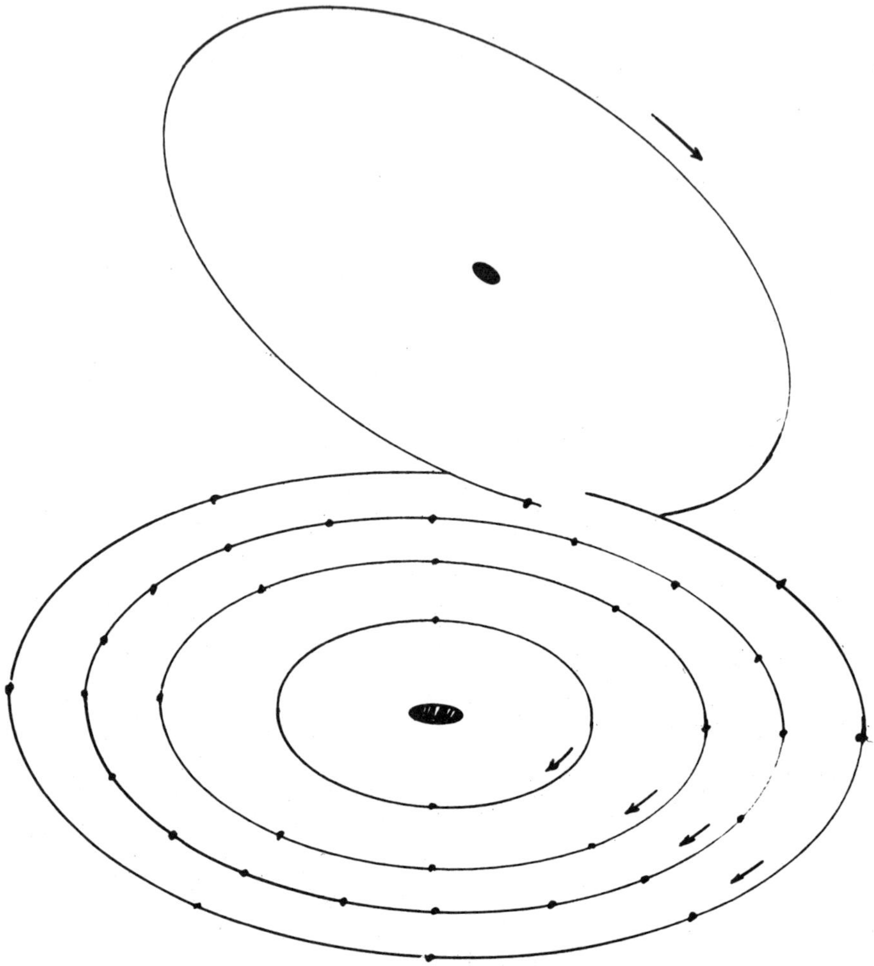

C. Scale diagram of HBr

Radius of Br: 0.65Å

Radius of H: 0.55Å

Z number Br: 35

Z number H: 1

point in that electron "cloud." In thinking of instantaneous moments, the electron is not a cloud, but a particle, and it has a specific, discrete location. These diagrams show the locations of electrons in *some* orbital path, instantaneously. It must be borne in mind, however, that all the electrons in any orbital path need not be, and probably are not, in the *same* orbital path—thus accounting for different orbital energies, different orbital frequencies, and different spectral lines.

of two chlorine atoms to breach the outer electron orbit of the iron atom. Ferric chloride would be the result of an electron from each of the outer orbits of three chlorine atoms to breach the outer electron orbit of the iron atom (see Fig. 15A and 15B).

In the case of nonpolar bonds, such as CH_4, each of the four hydrogen protons could enter into the *electronic* orbit of the carbon atom, with each hydrogen electron orbiting its own nucleus (see Fig. 16A and 16B).

As an interesting aside, it should be pointed out that the planet Jupiter is considered, by some astronomers, to be a small star, since it emits more energy than it receives from the sun. If this should be so, and it could be, it could be similar to a magnesium nucleus, with its 12 electrons (Jupiter has 12 moons) in the satellite orbit of the sun, which has 8 satellites (oxygen has 8 electrons), yielding, on a macroscopic scale, a molecule of MgO. Our count of 8 solar satellites is *not* an error, since we have pointed out earlier that we have elected to consider Pluto to be in a different system.

To continue the above parallel, it has been stated earlier that in the ambient emag sea all of the electrons around a nucleus ought to be, essentially, in the same plane, just as the planets of the solar system, with the exception of Pluto, are in essentially the same plane—the ecliptic. It should be noted, however, that they need not be, and are not in *exactly* the same plane. Also, as we consider Pluto to be in a different system, so, too, can different atoms in the same molecule be in different ambient emag seas, and, therefore, not in the same essential plane.

Jupiter, which is 5.2 AU from the sun, approximately 500,000,000 miles (483.6 million miles), has satellites at a maximum distance of 23,700,000 Km., or 14,722,000 miles. We see, therefore, that the farthest Jupiter satellite is an insignificant distance from Jupiter's nearest major solar planet, Mars, relatively. Alternating between our nuclear and solar planetary systems, while each planetary system (atom) is in its own plane, two associated planetary systems (molecule) need not be. However, there is no requirement that there be symmetrical, three-dimensional distribution of a multinuclear molecule, though it could happen.

The current scheme generally used in non-polar formula writing is still applicable to the description offered by the emag hypothesis, provided we understand that when we use a dash as the symbol of a bond, we are really indicating the intertwining of electron orbits between two atoms. For example, H_3PO_4 can still be written as

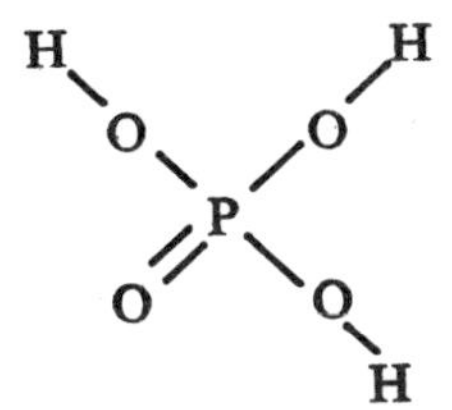

Figure 16. Possible spatial relationships of two nonpolar bonds

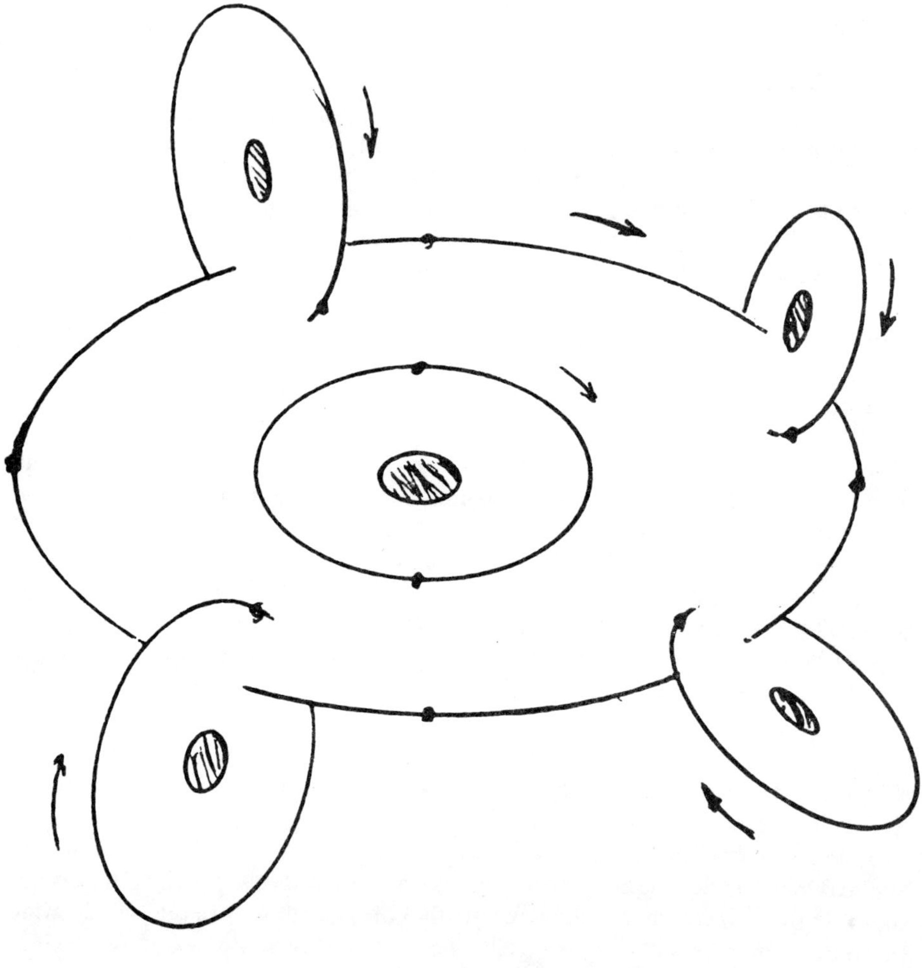

A. One possible version of a nonscale diagram of CH_4

Z number C 6
Z number H 1

Note hydrogen electron is in outer orbital ring of carbon.

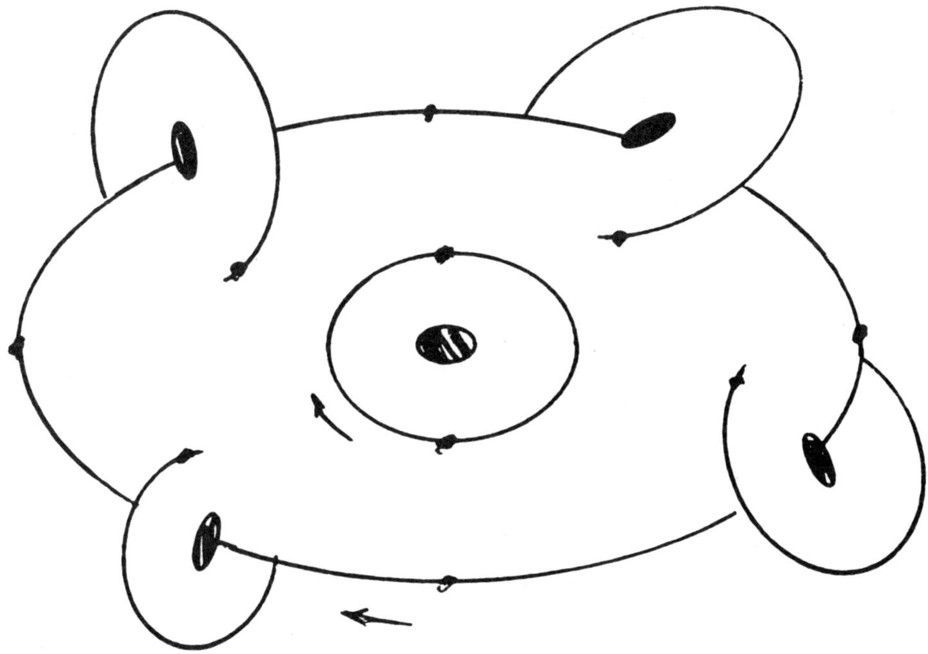

B. One possible version of nonscale diagram of CH_4

Z number C 6
Z number H 1

Note hydrogen proton is in outer electron orbit of carbon.

The bond does not indicate the sharing of electrons, which would be impossible and still allow orbital motion. However, the hydrogen atom of the OH group *can* mean that the hydrogen proton is in the electron orbit of the oxygen atom, and the hydrogen electron is orbiting the hydrogen nucleus, as the moon orbits the Earth, which orbits the sun (see Fig. 17A and B).

We have suggested, earlier in this chapter, that there can be, and probably are, numerous free electrons in the ambient emag sea, which have been knocked out of orbit by chance collisions. While some of these free electrons are available for capture by other atoms, this, too, is a chance phenomenon. It is the loss or the acquisition of such free electrons which in present thought is believed to give ions their charges, and to permit them to migrate in electric "fields," which truly determine that ions will migrate. However, the migration is not, according to the emag hypothesis, due to the charge, but rather to the change in positive or negative mass of the ion particle, which now becomes a different target for *other emag* currents of the electric "field."

Another ramification of the ion concept is of significance. Each time an atom becomes an ion, by gaining or losing an electron, the character of the atom must change, chemically, to a new element. We have seen, and agreed, that the chemical likeness of atoms is determined by the number of electrons in orbit. It is, therefore, quite obvious that an ion is a new element. However, after an ionic, chemical transaction is completed, the original number of electrons is reinstated in the atoms, and the old elements are regained. Such a state of affairs does not occur with reactions involving nonpolar bonds, unless, as in the case of CH_4, above, the hydrogen protons enter the *electronic* orbit of the carbon atom, as previously described. In other words, things may not be as they seem. The analytic components of a compound may be the elements with which we are familiar, but the synthetic component of a compound is an entirely different commodity. It is such differences that give physical and chemical characteristics to chemical substances.

Figure 17. Possible spatial relationships of H_3PO_4

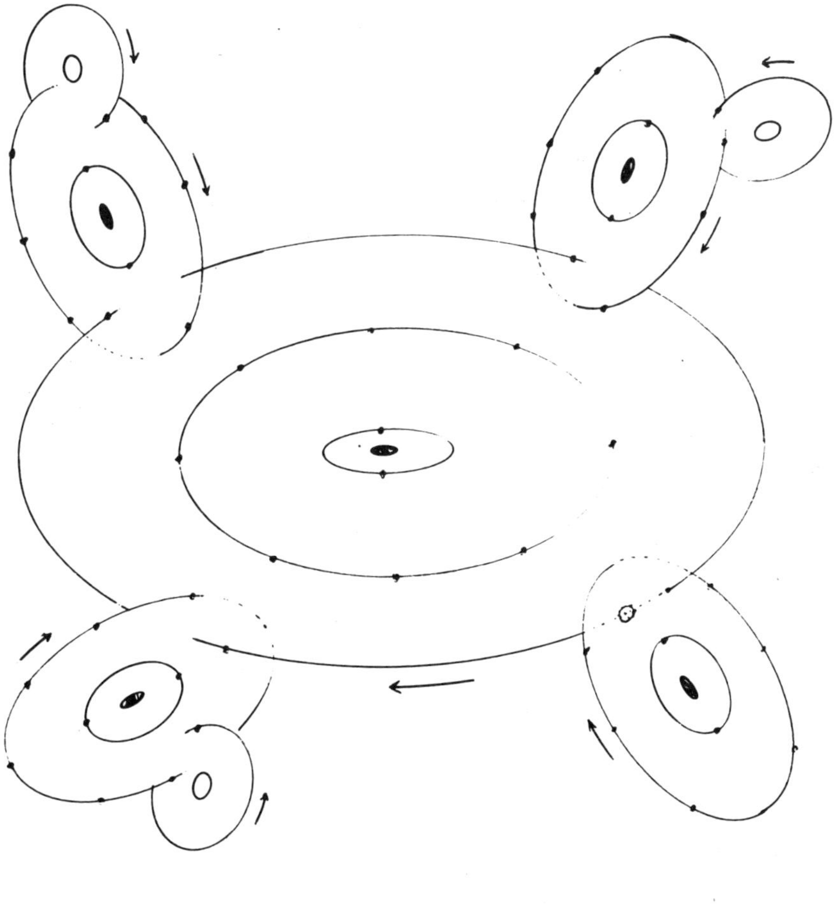

A. One possible version of nonscale diagram of H_3PO_4

Z number *P* 15
Z number *O* 8
Z number *H* 1

Note hydrogen electron is in outer orbital ring of oxygen.

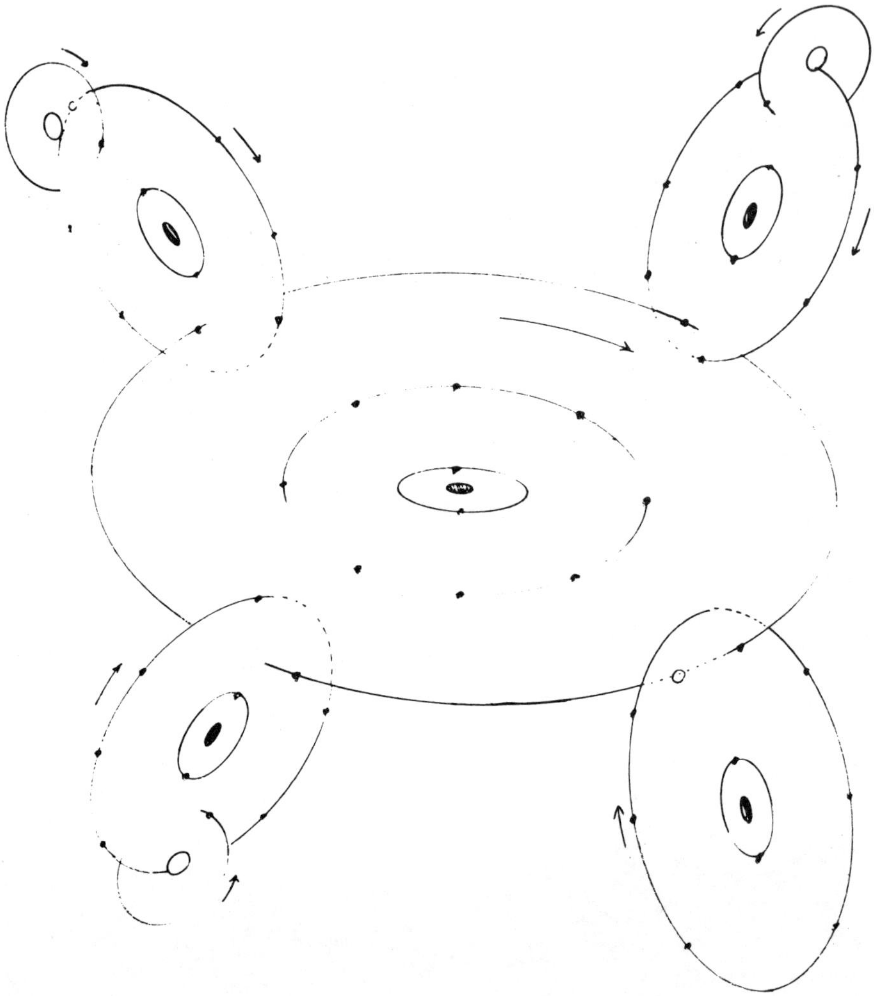

B. One possible version of nonscale diagram of H_3PO_4

Z number P: 15
Z number O: 8
Z number H: 1

Note hydrogen proton is in outer electron orbit of oxygen.

19. Batteries

In discussing ions, we indicated that the migration of ions in an electric "field" is not due to an attractive force nor an "action at a distance" repulsive force, but rather to the direct action collision force between the ion and the current of *other emags* pouring from the electrodes. Now let us reverse the procedure and discuss the action of an electrolyte into which are dipped two "conducting" plates—a battery.

With the proper selection of an electrolyte and the two plates, and remembering that there is no such thing as the absence of motion, there is set up an interaction and collisions between any and all phases. The collisions that occur between the moving molecules of the electrolyte and the moving molecules of the electrodes result in the dislodgement of electrons. Additionally, the general millieu—solvent, solute and plates—are permeated by the ever-present emag sea. Such a sea, as all seas with which we are familiar, have dominant and lesser currents and motions. In the case of the emag sea, however, we must also remember that it is composed of matter and antimatter—emags and *other emags*—to various degrees. Consequently, disruption of atoms, to a greater or lesser degree, occurs both by atom-atom collision and emag-*other emag*-atom collision. The result is what we know as a *potential difference* between the electrode phase and the solvent-solute phase. Equilibrium is established, and when the two electrodes are connected through a galvanometer, *current* is seen to flow.

As has been indicated throughout this dissertation, in adapting experimental observations to the emag hypothesis, it has been necessary, at times, to abolish currently accepted notions. And now we are at a similar crossroad. In the last paragraph we spoke of an electric *current*. Present thought describes an electric current as a *flow of electrons*. In the light of the past thinking, this was a reasonable definition. However, when we stop to think that the present thought requires either electron migration through an electrode of a battery, or at least serial displacement, along the electrode, of one electron by another, the idea of electron flow no longer seems reasonable. It appears far more reasonable to consider an electric current to be a flow of *other emags,* since this process *can* be almost instantaneous through the maze of atoms of a solid body or a solution. Consequently, we shall now redefine an *electric current* as a *flow of other emags.* This has the additional advantage that it is completely compatible with our previous description of an electric "field" as being a flow of *other emags.* What is the difference between an electric current and an electric "field"? The difference is that a *current is directed through a specific channel,* either a wire or an electrode or a solution, while a "field" is not *specifically* directed.

Getting back to our battery, we now know how the current is generated at the electrode-electrolyte junction, and how it travels through a conductor from one electrode to the other. We have already described, in the last section, how the migration of ions occurs in an electrolyte.

20. Magnetism and Magnetic "Fields"

We have discussed gravitational "fields" and electric "fields," but we have not yet discussed magnetic "fields." As usual, we shall utilize the concept of parallelism. Gravitational "fields" are the effects of the greater emag sea of the universe. Acting through such great distances, the magnitude of the force is relatively small. Electric "fields" are the effects of localized *other emag* currents which, in the past, have given the impression of charged bodies, rather than oppositely-massed bodies. Magnetic "fields" are in between. Magnetic effects are more local than gravitational "fields" and less local than electric "fields," and such, too, are the orders of magnitudes of these forces. The magnitudes are dependent upon the *flux* or concentrations of *other emag* effects per unit area. Just as we described the ion motions in an electric "field" as the resultant effect of *other emags,* so do we describe magnetic forces.

It appears that the positive mass-negative mass of electric and magnetic "fields" are of overriding importance. How do they work? Apparently the effect is due to differential permeabilities of substances. The internal structures of some substances tend to filter out either emags or *other emags,* preferentially. As implied earlier, in the discussion of mass defect, the internal structures of cobalt, iron, and nickel, with their maximal mass defects per baryon, must be very conducive to the preferential filtering process described as *ferromagnetism.* Now we must distinguish between so-called positive and negative magnetic poles.

Once again, we return to empiricism. It has been found, using the right hand rule, that if along a wire which is conducting an electric current (a stream of *other emags*), we point the thumb of the right hand in the direction of the current (which, prior to the emag hypothesis, was considered to flow from anode to cathode, contrary to the accepted, presumed flow of electrons), and if the fingers of the right hand point in the direction of a magnetic "field" in which the electric wire is suspended, the wire will move in the direction that the palm is facing, as depicted directly below in Fig. 18.

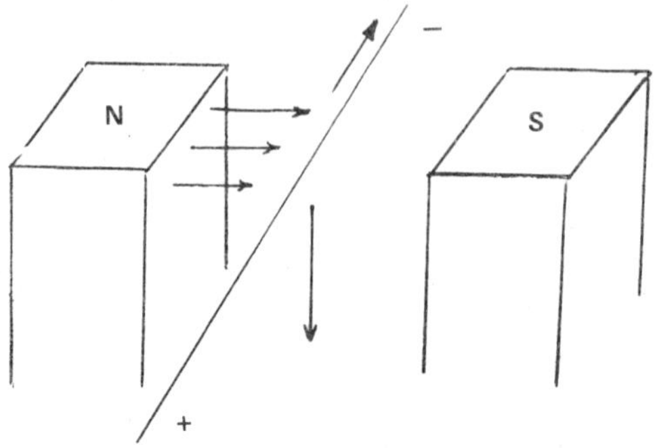

Figure 18. Effects of magnetic "fields"

This motion, of course, is due to the effect of the magnetic "field" upon the magnetic "field" set up around the wire, as the electric current flows in the wire. Since, in magnetism, empirically, the north or positive pole repels a like pole (when the wire moves down, for example), it is because it is being pushed down from above or "pushed" down from below.

How does a flow of *other emags* in a wire produce an encircling magnetic "field"? Clearly, *other emags* must escape the wire, perpendicularly, or we would not have a magnetic circulation around the wire. So we *must* conclude that the stream of *other emags* constituting the electric current is not confined to the wire, but is only *conducted* or *concentrated* or *guided* by the wire. Within the wire, by collision and/or filtration, some *other emags* are thrust out of the wire, at right angles to it. But is this sufficient? If this were the whole story, these *other emags* would fly out into space and there would be no magnetic *circulation around the wire*. However, it would appear that the circular magnetic "field" is due to the collisions resulting from the escaping *other emags* with the surrounding *other emags* of the magnetic "field." In the absence of a surrounding magnetic "field" of our own choosing, the above effect must be due to the Earth's magnetic "field."

This raises another concept. Unfortunately, convention prior to the emag hypothesis defined the direction of an electric current as moving from "higher potential to lower potential," or from anode to cathode–this despite the fact that the current was considered to be a flow of negatively charged particles: electrons. This confusion was compounded by the convention that the direction of a magnet's magnetic "field" was defined as going from the north, or positive, pole to the south, or negative, pole. The above would appear to be consistent except, when we stop to consider that a magnet's north pole is defined as the pole pointing to the Earth's north pole–which, of course, makes the magnet's north pole a south pole! Can we reconcile these differences with the emag hypothesis? Figure 19 shows the usual delineation of magnetic "lines of force."

In making changes in a concept, it generally becomes necessary to make supplementary changes. Consequently, Figure 20A indicates a different conception of "lines of force." Consider a bar magnet to be the equivalent of an electric coil. As is well known, each turn of a coil becomes additive to the neighboring turns, and results in a central magnetic "field" (Fig. 20B.) The emag hypothesis states that the magnetic effects are similar to electric effects—that is, *other emag* flows—except that the flux concentrations per unit of cross sectional area are less.[1] When the *other emag* flows are in the same direction, we have *constructive interference*. When the *other emag* flows are in opposite directions, we have *destructive interference*. Now we are in a position to explain, without "action at a distance," the effects of magnetism.

Figure 20 is a schematic representation of magnetic force. The emag hypothesis proposes that ferromagnetic substances are preferentially permeable to *other emags*. Since *other emags* are of a greater population in the ambient emag sea, any relative increase in *other emag* concentrations, with their negative mass, results in a greater push where that *other emag* concentration is greater.[2]

Consequently, the effect is similar to a *pull,* because the ambient emag sea becomes relatively greater, negatively. According to the emag hypothesis, the south pole of a magnet is the pole of greater *other emag* concentration exit. The north pole, correspondingly, is the pole that is receiving the inflow of *other emags*. What makes the flow of *other emags* exit at the south pole? Nothing in particular. The pole from which the *other emags* exit is found to correspond to our definition of the magnetic south pole. However, it must be pointed out, just as we did when we changed the convention of electrons to negative mass, rather than negative charge, and protons and positrons to positive mass, rather than positive charge, that we are obliged to recognize that the usually titled north magnetic pole is really a south pole, or north-seeking pole.

Why should magnetic substances have greater permeability to *other emags?* It would appear that the alignment of such atoms, perhaps externally, but surely internally, permits such preferential permeability. Let us look to the atomic structures of magnetic elements, to see if we have an answer. Returning to our previous discussion of mass defect, we noted that the second maximum of the curve of mass defect per baryon was at atomic mass 56—iron. Adjacent to iron are cobalt and nickel on one side, and manganese on

1. It would appear that there may well be other universes where the *other emag* is the dominant particle, and where the emag is the lesser particle, and an electric current would be the flow of emags.

2. The reason the emag sea is more abundant in *other emags* than emags is that we have elected to consider the basic matter of our universe, namely the proton, to be of positive mass. Since, initially, the number of emags and *other emags* were equal, in our universe, the residual *other emags* dominate the emag sea—and of course, the electrons.

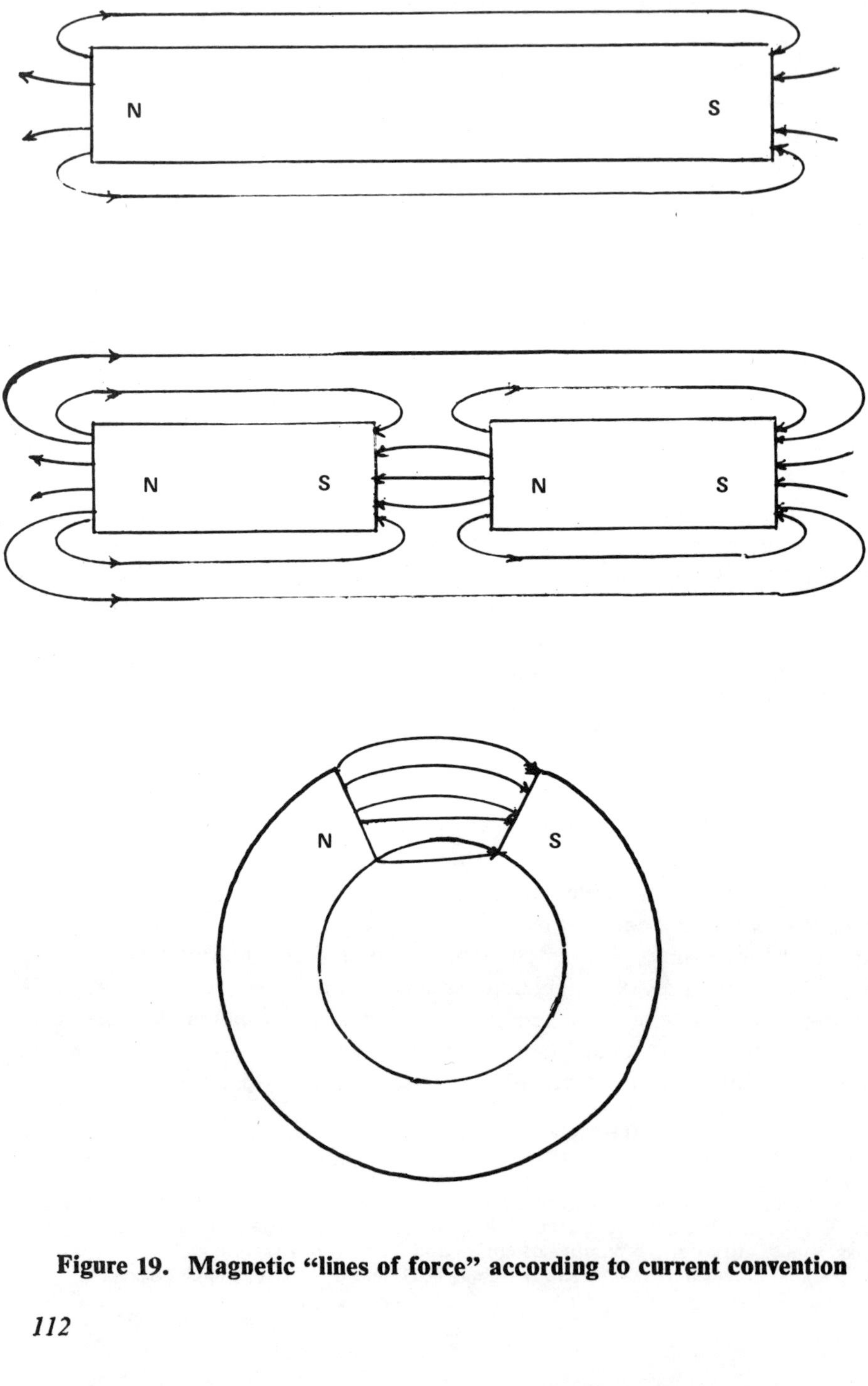

Figure 19. Magnetic "lines of force" according to current convention

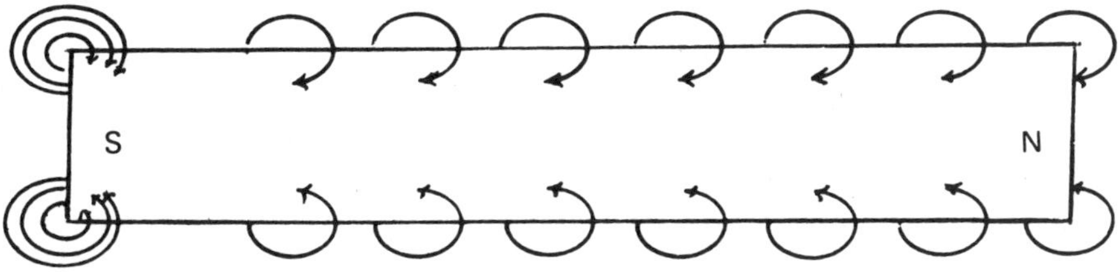

A

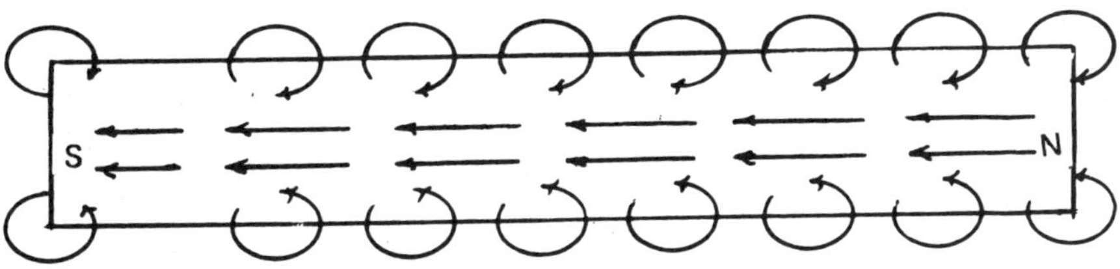

B

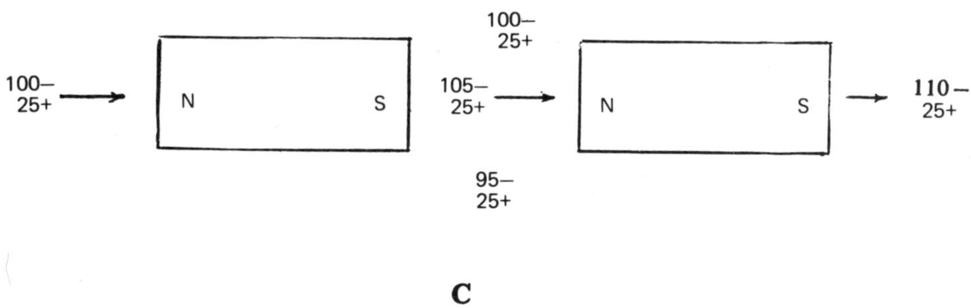

C

Figure 20. Magnetic force according to the emag hypothesis

the other side. Iron, cobalt, and nickel are the ferromagnetic elements of prominence. While copper and manganese are only slightly ferromagnetic, alloys of copper, manganese, and aluminum form very magnetic materials (see Tables 5A and 5B).

In order to account for the greater permeability of magnetic elements, it would appear that resistance to the flow of *other emags* needs to be reduced. Looking, again, at the curve of mass defect per baryon, we conclude that the reason the mass defect exists at all is that the *other emags* present in the atomic nucleus, by virtue of their negative mass, reduce the total mass of the nucleus. The greater number of *other emags* in such a nucleus must result in winding and intertwining channels of zero mass within the substance of the nucleus. These channels, evidently, provide relatively unhindered access through the nucleus to the *other emags* of the ambient emag sea. In other words, there is a "sponge effect." Sponges are soft and filled with much air. Therefore, the specific gravity of the sponge is relatively small. Should not a ferromagnetic nucleus, by virtue of the above analysis, have a relatively smaller specific gravity? We should expect so.

Tables 5A and 5B show the breakdown of the magnetic elements, comparable to Tables 3A and 3B. Note, in particular, the almost consecutive and maximal mass defects per baryon of the elements. Bismuth, which is paramagnetic, is included for comparison.

Table 5A

1	2	3	4	5	6	7	8	9	10
#	Z#	Element	Handbook mass if $_6C^{12}$ is 12.00000AMU	*N* 1.00865 AMU	*P* 1.007277 AMU	*E* 5.48327 × 10^{-4} AMU	Calculated mass by parts AMU	Mass defect AMU	Mass defect/ baryon AMU
1	13	$_{13}Al^{27}$	26.9815	14	13	13	27.2228	.2413	.00894
2	25	$_{25}Mn^{55}$	54.9381	30	25	25	55.4551	.5170	.00940
3	26	$_{26}Fe^{56}$	55.9349	30	26	26	56.4630	.5280	·00943
4	27	$_{27}Co^{59}$	58.9332	32	27	27	59.4881	.5549	.00940
5	28	$_{28}Ni^{58}$	57.9353	30	28	28	58.4786	.5433	.00937
6	29	$_{29}Cu^{63}$	62.9298	34	29	29	63.5210	.5912	.00938
7	83	$_{83}Bi^{209}$	208.980	126	83	83	210.739	1.759	.00842

Table 5B

1	2	3	4	5	6	7	8	9	10
#	Z#	Element	Calculated mass from Hdbk if $_1H^1$ is 1 AMU		P 1.00054 AMU	$-E$ 5.44662 × 10^{-4} AMU	Calculated mass by parts AMU	Mass defect AMU	Mass defect/ baryon AMU
1	13	$_{13}Al^{27}$	26.7720		27	13	27.0075	.2355	.00872
2	25	$_{25}Mn^{55}$	54.5115		55	25	55.0161	.5046	.00917
3	26	$_{26}Fe^{56}$	55.5006		56	26	56.0161	.5155	.00920
4	27	$_{27}Co^{59}$	58.4756		59	27	59.0172	.5416	.00918
5	28	$_{28}Ni^{58}$	57.4855		58	28	58.0161	.5306	.00915
6	29	$_{29}Cu^{63}$	62.4412		63	29	63.0182	.5770	.00916
7	83	$_{83}Bi^{209}$	207.357		209	83	209.068	1.711	.00819

Let us calculate the volumes of these atoms. We shall do it in the following manner, using data from Tables 6A and 6B.

Using Table 6B:

1. From the specific gravity of an element, we calculate the mass of 1 cc—column #4.
2. Dividing the mass per cc by the gram atomic mass, gives us the number of moles per cc.
3. Column #5 gives us the number of atoms per cc, which is derived by multiplying the number of moles per cc, calculated in #2 above, by Avogadro's number-6.022 × 10^{23}.
4. Dividing the number of atoms per cc into 1 cc, or 10^{-6} m^3, gives us the volume occupied by a single atom, including its space of motion. This is column #6. If we assume the above volume and space of motion of the atom to be a sphere, which it is not, we can come up with a figure for the radius of that sphere, which is column #7.
5. Column #8 gives a number representing the radius of column #7 per proton of the nucleus.

Using Table 6A:

1. From the spectral lines of the elements, column #3, we calculate the apparent frequency of these lines, column #5, add to that frequency the frequency of the "stationary orbit," and with the resultant real frequency, column #6, we calculate the real radius of the elements, using Kepler's third law: column #7

2. The real radius which we have just calculated is referable to the flat planetary system we have discussed previously. However, for purposes of comparison with Table 6B, we will use that radius to calculate the volume of a sphere: column #8.
3. Column #10 shows the volume of that sphere per proton of the nucleus.
4. Column #11 shows the radius of column #7, per proton of the nucleus, and
5. Column #12 shows the radius of column #7, per orbital electron.

Column #9 of Table 6A shows the difference between the real radius (flat) of column #7 in Table 6A and the calculated radius of column #7 in Table 6B, which was dependent upon the assumed spherical volume of the atom plus its space of motion. We see, with some exceptions, that this reaches a minimum in the region of the ferromagnetic elements. In other words, the radii of the ferromagnetic elements are closer to the larger radii of the calculated radii of the atoms plus their spaces of motion. This is exactly what we would expect with a less massive of "spongier" nucleus. To document this concept, we should note that the most massive elements have relatively smaller radii than the less massive do. The composite Figure 21 shows many of these relationships. And Newton's law also tells us that this should be the case. Newton's law says that the force between two bodies is proportional to the product of their masses and inversely proportional to the square of the distance between them. Thus, if the force remains constant, and the mass of one of the bodies remains constant (the orbital electron, for example), as the mass of the other body increases (the nucleus, for example), the distance between the two bodies decreases.

The idea of "filtration" may seem contrived, but under the pressure of the emag sea, which is composed of *other emags* in apparent excess over emags, a filtration process is not unreasonable. Comparison with the filtration of water through a filter paper, without apparent alteration of the filtrate, is similar, even if the filter paper *is* made wet. Comparison with the filtration of mercury through leather is similar, and the leather does not even get wet.

The multiple curves in the graph that is Figure 21 seem complicated and, though they are not, they do require an extended explanation. All the curves, except where indicated otherwise, use data dependent upon the emag hypothesis.

1. Curve #1 is the relationship between the atomic mass number (or in some cases, the atomic mass in AMUs, where appropriate) on the abscissa and the mass defect per atom on the ordinate. The curve is almost a straight line progression. The mass defect per atom is a very small percentage of the atomic mass. These mass defects are found in Table 3B, column #8, assuming ${}_1H^1$ to be 1

AMU and evaluating mass by parts according to the emag hypothesis, i.e., no neutrons, all baryons are protons, and all orbiting electrons are of negative mass.[1] The comparable curve for mass defect using accepted data is grossly very similar.

2. Curve #2 shows the mass defect per baryon, which is roughly equivalent to the mass number. In this curve, however, all baryons are assumed to be protons, since there are no neutrons, according to the emag hypothesis. The curve shows a maximum at the ferromagnetic elements, iron, cobalt, and nickel. See Table 3B, column #10.

In an effort to find additional correlations between the ferromagnetic elements and other atomic parameters, the following data were plotted:

3. Curve #3 shows the maximal and minimal real radii, r_r, as calculated from spectral lines, using Kepler's formula. See Table 6A, column #7. The curve is marked with S's.
4. Curve #4 shows the plot of the real radius, r_r, versus mass number. See Table 6A, column #11. This curve also shows maxima and minima, and is hatched from southwest to northeast.
5. Curve #5 shows the plot of the real radius, r_r, per orbital electron, or $Z\#$, versus the mass number. See Table 6A, column #12. Also see the last footnote.
6. Curve #6 shows r_a, the apparent radius of the atoms including their space of motion, as determined for Table 6B. See column #7 of that table.
7. Curve #7 shows the plot of the difference between the apparent radius, r_a, of curve #6 and the real radius, r_r, of curve #3. This curve also plots the maxima and minima, and is hatched from northwest to southeast. See Table 6B, column #9.

The composite graph (Fig. 21) shows several interesting correlations. Curve #3 shows a minimal difference between the maximal and minimal real radii of the ferromagnetic elements. In other words, the electron orbits would appear to be very stable and almost in fixed orbits. Confirmation of this relationship appears in Curve #7. No other elements in the graph show this unique correlation.

1. We have been assuming that all the orbital atomic bodies are electrons with negative mass. Regrettably, this may not be the case. We have reached the first conclusion, naturally, because in the *time of flight* experiment, with hydrogen, we *have* demonstrated the electron to be of other kind than the proton. However, the abolition of "action at a distance" and of the magical nature of "charges" does not make a "neutral" atom a requirement, and, as mentioned above, may not even be the case. Of course, if this should be so, all of our calculations are in error until we can determine how many orbital bodies in atoms are electrons and how many are positrons.

It is to be expected, of course, that r_r calculated from spectral lines should be less than r_a, representing the radius of the sphere containing the atom plus its space of motion. Why this should not be so in the case of zinc is unexplained at this time.

We must bear in mind, whether we are dealing with so-called stable atoms with large mass defects per baryon, unstable atoms of radioactive elements with small mass defects per baryon, gravitational, magnetic, or electrical effects, electrolytes, batteries, or chemical bonding of the various types, that there is a common denominator. That common denominator is the interplay of emags and *other emags* and their aggregates with the ubiquitous emag sea. We must remember that even in those interplays that appear stable to us there is really a "constant," dynamic equilibrium. *There is constant motion!*

Table 6A

1	2	3	4	5	6	7	8	9	10	11	12
#	Element	Spectral line Å	Calc. M from HDBK if $_1H^1$=1AMU	$f_a = c/\lambda$ × 10^{14}	$f_r = f_a + f_o$ $f_o = .4950$ × 10^{14}	$r_r = f_r^{-2/3}$ Å	Vol (if sph) $m^3 \times 10^{-29}$	r Table B − r_r Å	Vol/prot (if sph) $m^3 \times 10^{-29}$	r_r/prot	r_r/E
1	$_3Li^7$	3232	6.96154	9.2757	9.7707	1.0156	.56608	.7133	.0808	.1451	
		4603		6.5129	7.0079	1.2675	.85297	.4614	.12185	.1811	
		6103		4.9122	5.4072	1.5067	1.4327	.2222	.20467	.2152	
		6707		4.4698	4.9648	1.5949	1.6994	.1340	.24277	.2278	.5316
2	$_4Be^9$	2348	8.94222	12.768	13.262	.82840	.23816	.4141	.02646	.0920	
		2650		11.313	11.808	.89512	.30042	.3474	.03338	.0994	
		3321		9.0271	9.5221	1.0332	.46200	.2093	.05133	.1148	.2583
3	$_6C^{12}$	2478	11.9068	12.098	12.593	.85753	.26414	.4194	.02201	.0715	.1429
4	$_{11}Na^{23}$	3302	22.8113	9.0790	9.5740	1.0294	.45692	1.074	.01987	.0448	
		5688		5.2706	5.7656	1.4436	1.2602	.6600	.05479	.0628	
		5895		5.0855	5.5805	1.4753	1.3450	.6283	.05848	.0641	.1341
5	$_{13}Al^{27}$	3082	26.7720	9.7271	10.222	.98547	.40088	.5929	.01485	.0365	
		3092		9.6957	10.191	.98747	.40333	.5909	.01494	.0366	
		3944		7.6012	8.0962	1.1512	.63906	.4272	.02778	.0426	
		3961		7.5685	8.0635	1.1543	.64434	.4241	.02386	.0428	.0888
6	$_{19}K^{39}$	4044	38.6612	7.4132	7.9082	1.1694	.66985	1.441	.01718	.0300	
		7698		3.8944	4.3894	1.7314	2.1741	.8786	.05575	.0444	.0911
7	$_{25}Mn^{55}$	4030 ↓	54.5115	7.4390	7.9340	1.1668	.66539	.2750	.01210	.0212	
		4034		7.4316	7.9266	1.1676	.66676	.2742	.01212	.0212	.0467
8	$_{26}Fe^{56}$	3581	55.5006	8.3717	8.8667	1.0835	.53281	.3250	.00941	.0193	
		3719 ↓		8.0610	8.5560	1.1096	.57225	.2989	.01022	.0198	
		3748		7.9987	8.4937	1.1150	.58065	.2935	.01037	.0199	.0429

NOTE: Arrows in spectral line column indicate a range.

Table 6A (contd.)

1	2	3	4	5	6	7	8	9	10	11	12
#	Element	Spectral line Å	Calc. M from HDBK if $_1H^1$=1AMU	$f_a = c/\lambda$ × 10^{14}	$f_r = f_a + f_o$ $f_o = .4950$ × 10^{14}	$r_r = f_r^{-2/3}$ Å	Vol (if sph) $m^3 \times 10^{-29}$	r Table B $- r_r$ Å	Vol/prot (if sph) $m^3 \times 10^{-29}$	r_r/prot	r_r/E
9	$_{27}Co^{59}$	3453	58.4756	8.6820	9.1770	1.0589	.49734	.3170	.00843	.0179	
		3465		8.6519	9.1469	1.0612	.50059	.3147	.00848	.0180	
		3579		8.4950	8.9900	1.0736	.51834	.3023	.00879	.0182	.0398
10	$_{28}Ni^{58}$	3524	57.4855	8.5071	9.0021	1.0726	.51690	.2954	.00891	.0185	.0383
		3414		8.7812	9.2762	1.0514	.48685	.3166	.00839	.0181	
11	$_{29}Cu^{63}$	3247	62.4412	9.2328	9.7278	1.0186	.44269	.3846	.00727	.0162	
		5153		5.8178	6.3128	1.3589	1.0511	.0443	.01668	.0216	
		5218		5.7453	6.2403	1.3694	1.0757	.0338	.01707	.0217	.0472
12	$_{30}Zn^{64}$	2138	63.4327	14.022	14.517	.77998	.19876	.7420	.00311	.0122	
		3300		9.0845	9.5795	1.0290	.45639	.4930	.00713	.0161	
		4810		6.2326	6.7276	1.3024	.92538	.2196	.01446	.0204	
		6362		4.7122	5.2072	1.5450	1.5448	−.0230	.02414	.0241	.0515
13	$_{35}Br^{80}$	1540	78.3056	19.467	19.962	.63076	.10512	1.5661	.00131	.0079	
		1633		18.358	18.853	.65526	.11785	1.5416	.00147	.0082	.0187
14	$_{47}Ag^{107}$	3280	106.074	9.1399	9.6349	1.0257	.45122	.5623	.00422	.0096	
		3382		8.8643	9.3593	1.0451	.47815	.5429	.00447	.0098	
		5209		5.755	6.2500	1.3680	1.0724	.2200	.01002	.0128	
		5465		5.4856	5.9806	1.4088	1.1712	.1792	.01094	.0132	.0300
15	$_{51}Sb^{121}$	2068 ↘	119.965	14.497	14.992	.76341	.18636	1.159	.00154	.0063	
		3267 ↙		9.1763	9.6713	1.0225	.44779	.9002	.00370	.0084	.0200
16	$_{78}Pt^{195}$	2659 ↘	193.451	11.274	11.769	.89710	.30242	.6320	.00460	.0046	
		3064 ↙		9.7843	10.279	.98182	.39645	.5473	.00203	.0050	.0126

Note: Arrows in spectral line column indicate a range.

Table 6A (contd.)

1	2	3	4	5	6	7	8	9	10	11	12
#	Element	Spectral line Å	Calc. M from HDBK if $_1H^1$=1AMU	$f_a \doteq c/\lambda$ $\times 10^{14}$	$f_r = f_a + f_o$ $f_o = .4950$ $\times 10^{14}$	$r_r = f_r^{-2/3}$ Å	Vol (if sph) $m^3 \times 10^{-29}$	r Table B $- r_r$ Å	Vol/prot (if sph) $m^3 \times 10^{-29}$	r_r/prot	r_r/E
17	$_{79}Au^{197}$	2427	195.438	12.352	12.847	.84618	.25379	.7426	.00129	.0043	
		2675		11.207	11.702	.90052	.30589	.6883	.00155	.0046	
		2802		10.699	11.194	.92756	.33428	.66124	.00170	.0047	.0117
18	$_{80}Hg^{202}$	1849	198.415	16.214	16.709	.71018	.15004	1.087	.00074	.0035	
		2536		11.821	12.316	.87034	.27616	.9271	.00087	.0043	
		3650 ↓		8.2134	8.7084	1.0966	.55237	.7008	.00273	.0054	
		3654		8.2044	8.6994	1.0973	.55343	.7001	.00274	.0054	.0137
19	$_{82}Pb^{204}$	2169	202.389	13.822	14.317	.78722	.20435	1.1320	.00100	.0038	
		2614		11.469	11.964	.88732	.29264	1.032	.00143	.0044	
		2833		10.582	11.077	.93408	.34138	.9851	.00167	.0046	
		3683		8.1398	8.6348	1.0128	.56180	.8164	.00275	.0054	
		4057		7.3894	7.8844	1.1717	.67381	.7475	.00330	.0057	.0143
20	$_{83}Bi^{209}$	2061	207.357	14.546	15.041	.76176	.18516	1.274	.00088	.0036	
		2276		13.172	13.667	.81199	.22425	1.224	.00107	.0039	
		2780		10.784	11.279	.92290	.32927	1.113	.00158	.0044	
		2809		10.672	11.167	.92906	.33591	1.106	.00161	.0044	
		2891		10.370	10.865	.94619	.35483	1.089	.00170	.0045	
		2938		10.204	10.699	.95596	.36594	1.080	.00175	.0046	
		2989		10.030	10.525	.96646	.37813	1.069	.00181	.0046	
		3067		9.7747	10.270	.98240	.39715	1.053	.00190	.0047	
		4722		6.3488	6.8438	1.2877	.89440	.7478	.00428	.0062	.0155

NOTE: Arrows in spectral line column indicate a range.

Table 6B

1	2	3	4	5	6	7	8
#	Element	Calc. Mass From HDBK if $_1H^1$ is 1 AMU	Sp' Gr.	# atoms per cc × 10^{23}	Vol 1 atom incl. space of motion r $m^3 \times 10^{-29}$	Radius of sph. cont. atom + its space of motion Å	Radius of sphere cont. atom + its space of motion per proton: Å
1	$_3Li^7$	6.96154	.534	.46193	2.1648	1.7289	.2470
2	$_4Be^9$	8.94222	1.848	1.2445	.80354	1.2425	.1381
3	$_6C^{12}$	11.9068	2.2670	1.1466	.87214	1.2769	.1064
4	$_{11}Na^{23}$	22.8013	.971	.25645	3.8994	2.1036	.0915
5	$_{13}Al^{27}$	26.7720	2.6989	.60708	1.6472	1.5784	.0584
6	$_{19}K^{39}$	38.6612	.862	.13427	7.4477	2.6100	.0669
7	$_{25}Mn^{55}$	54.5115	7.21	.79650	1.2555	1.4418	.0262
8	$_{26}Fe^{56}$	55.5006	7.874	.85436	1.1705	1.4085	.0252
9	$_{27}Co^{59}$	58.4756	8.9	.91655	1.0910	1.3759	.0233
10	$_{28}Ni^{58}$	57.4855	8.902	.93254	1.0723	1.3680	.0236
11	$_{29}Cu^{63}$	62.4412	8.96	.86413	1.1572	1.4032	.0223
12	$_{30}Zn^{64}$	63.4327	7.133	.67717	1.4767	1.5220	.0238
13	$_{35}Br^{80}$	78.3056	2.928	.22517	4.4411	2.1969	.0275
14	$_{47}Ag^{107}$	106.074	10.50	.59610	1.6776	1.5880	.0148
15	$_{51}Sb^{121}$	119.965	6.691	.33587	2.9773	1.9227	.0159
16	$_{78}Pt^{195}$	193.451	21.45	.66772	1.4976	1.5291	.0078
17	$_{79}Au^{197}$	195.438	19.32	.59530	1.6798	1.5888	.0081
18	$_{80}Hg^{202}$	198.415	13.546	.41113	2.4323	1.7974	.0089
19	$_{82}Pb^{204}$	202.389	11.35	.33771	2.9611	1.9192	.0094
20	$_{83}Bi^{209}$	207.357	9.747	.28307	3.5327	2.0355	.0097

21. *Refraction and the Emag Hypothesis*

Refraction of white light by a prism is a well-known phenomenon. Nonetheless, at this time, it needs to be analyzed, in depth. As Snell's Law indicates, the ratio of the sine of the angle of incidence, when light travels from one medium to another, to the sine of the angle of refraction, is constant for any two particular media. This ratio is defined as the index of refraction:

$$\frac{\text{sine } \theta_i}{\text{sine } \theta_r} = n$$

But what are the factors responsible for the phenomenon of refraction? Let us enumerate them and see what their effects are, if any:

1. The frequency of light in the new medium.
2. The length of the path in the new medium.
3. The momenta of the "photons" in the new medium.
4. The velocity of light in the new medium.

We know that when white light is refracted by a prism the result is a multicolored band, called the spectrum. It is also known that red light, with a lesser frequency, is refracted less than blue light, with a greater frequency. Our first approximation, therefore, is that the refraction of colored light varies, in some way, as the frequency, and this is true whether the light is going from the less dense medium to the more dense medium, or vice versa (see Figure 22).[1] Though the blue light of greater frequency is clearly refracted or *bent* more than the red light, note that the angles of incidence and

1. Cauchy's equation is $n = A + \frac{B}{\lambda^2} + \frac{C}{\lambda^4} + \ldots$

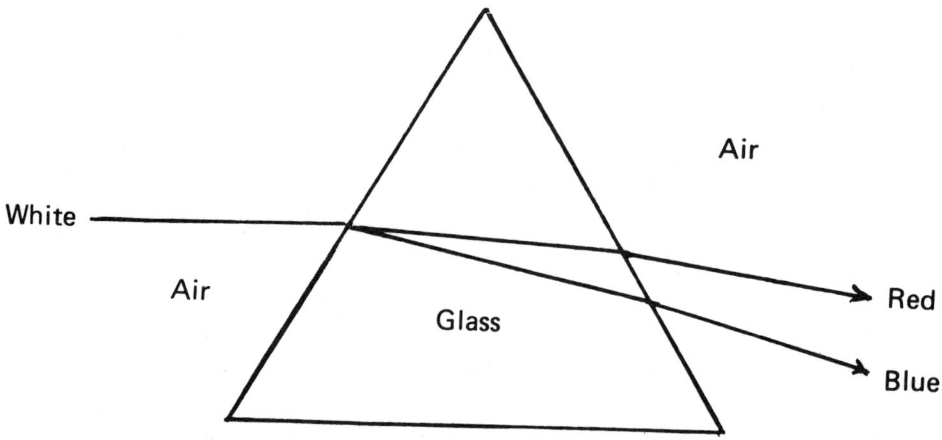

Figure 22. The refraction of white light by a glass prism

the angles of refraction, by convention, are measured from the *normal.*[2] So the *angle of refraction* of blue light, on entering a denser medium, is *less* than the *angle of refraction* of red light. Similarly, the *angle of refraction* of blue light on entering a less dense medium is *greater* than the *angle of refraction* of red light, on entering a less dense medium (see the last footnote). When either red light or blue light is used as the source, rather than white light, the angles of refraction are unchanged for each color. Laser light, of course, would give ultra-sharp refraction lines, as opposed to ordinary colored light, because the latter are mixed, close frequencies.

When white light is directed at a plate of glass, refraction occurs twice, and the exiting beam is parallel to the entering beam, and it is also white. (See Fig. 23.) Let us draw Figure 24 by modifying Figure 23. (See Figure 24.) It is clear that the white light entering the glass is refracted, differentially, so that the red light and the blue light are separated in the glass. On exiting the glass, however, the differential refraction is in the reverse direction, and the blue light and the red light exit in parallel beams, both parallel to the entering beam, and giving the effect of white light. These experiments indicate that not only are the frequencies of light important, but also the lengths of their paths. The lengths of the paths of red light and blue light, in traversing a prism, are significantly different from the lengths of their paths in traversing a plate of glass.

The next factor we need to consider is the momenta of the "photons" of red and blue light. We have mentioned earlier that "photons" have mass and it is this mass which is responsible for their momenta. We have also indicated that "photons" associated with different frequencies have different masses (see Chapter 9). As white light travels through air, the air molecules are so widely dispersed in this rare medium that the number of collisions occurring between light "photons" and air molecules, is relatively small. Additionally, due to the great velocity of light "photons," in these collisions it is the air molecules (of greater mass but of slower velocity, and, therefore, smaller momenta) that are dispersed, significantly, rather than the "photons." However, when these "photons" enter a denser medium, such as glass, though the glass molecules are of relatively slow velocity, their mass is quite great. In collisions with these molecules, therefore, the "photons" tend to be dispersed. This would result in diffused light, rather than a beam, were it not for the fact that a recoiling "photon," after collision with a heavy glass molecule, collides again with another glass molecule, in the opposite direction. The basic directional velocity of the original entering beam is altered, and compromised

2. It is regrettable that Snell elected to define the index of refraction as the ratio of the sines of the angles of incidence and refraction, using the *normal.* Had he chosen the *tangent*, as the point of reference, he would have arrived at the same index of refraction, *n*, by using the ratio of the cosines. This may not seem important, except that by such a maneuver, the index of refraction of a medium automatically becomes its own reciprocal on leaving the medium and on entering a new medium.

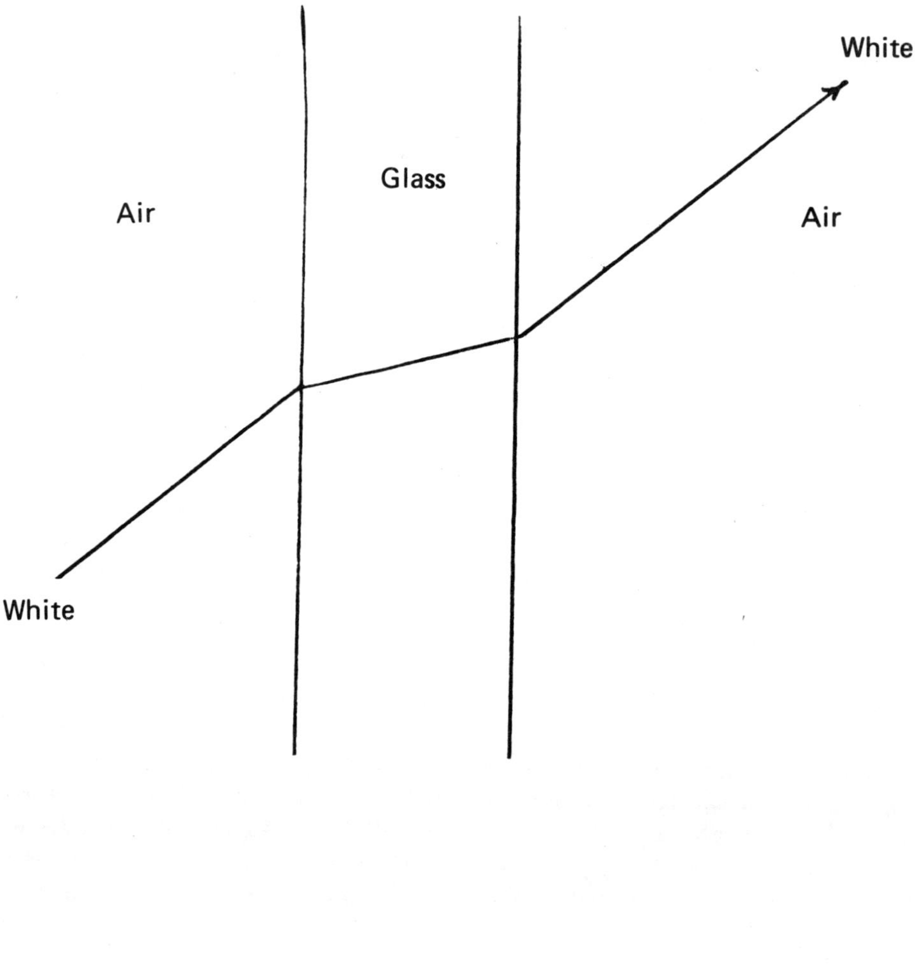

Figure 23. Double refraction of white light by a glass plate

CROWN GLASS

	Blue	Red
Index of Refraction	1.528	1.513

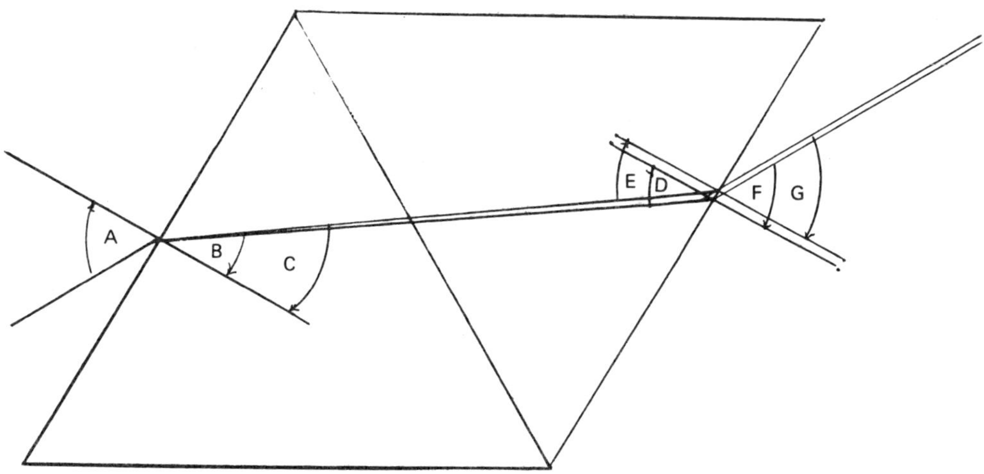

Figure 24. Scale drawing of Figure 23 (see text for explanation)

A = Angle of incidence = 60°
B = Angle of refraction, blue = 34.525°
C = Angle of refraction, red = 34.917°
D = Angle of incidence, blue, in glass = 34.525°
E = Angle of incidence, red, in glass = 34.917°
F = Angle of refraction, glass to air, blue = 60°
G = Angle of refraction, glass to air, red = 60°

momentarily, in the denser medium, but on exiting the glass into the air again, the direction of exit remains the direction of last contact with the glass.

The last factor to evaluate is the velocity of light. It is clear that red light is refracted less in glass than is blue light. It is also clear that the momentum of the red "photon" is less than the momentum of the blue "photon," if their velocities are the same. It is known that the velocity of light is less in water, for example, a denser medium, than in air, a sparser medium. So the velocities of red light and blue light are less in glass than in air. The question is, are they reduced by the same amount? It is quite clear that the phenomenon of refraction is completely due to the reduction of the velocity of light as it enters a denser medium from a less dense one. It also seems clear that the amount of reduction of the velocity varies, in some manner, as the frequency. In other words, blue light is bent more on entering a glass prism from air because the velocity of the blue light is reduced more than are the comparable parameters of red light. Similarly, the degree of increase of the velocity of blue light, as compared to the increase of the velocity of red light, as they both exit from a glass prism, is greater. It would appear that the greater slowing of blue light, and the greater refraction of blue light, are due to the greater number of collisions by the blue "photons" as opposed to the red "photons."

To summarize, then, the emag hypothesis states that the velocity of white light, in the emag sea (ether or *vacuo,* according to previous concepts) is dependent upon the momentum of orbiting electrons in various atoms, the energies of any particular orbital electron being transmitted to emags and emag aggregates in the emag sea. These emags and emag aggregates constitute "photons" which transmit the energy of electromagnetic radiation. The frequencies of these "photons" are dependent upon the frequencies of the orbiting electrons. From any particular electron source, since the mass and velocity of that electron are constant, at any particular time, the velocity of that particular "photon" will be constant (the velocity of light) and the mass of that "photon" will be constant. However, since electron velocities vary, depending upon the orbit in which they find themselves, the masses of the buffeted "photons" will also vary.

As a matter of practical observation, it appears that the velocities of the various colors are essentially the same (but probably not exactly) in the emag sea. When these various colors which constitute what the receptor (retina, for example) interprets as different colors enter air, their various "photons" of different frequencies are slowed minimally, so that we are not aware of the differential slowing. But when these same "photons" of various frequencies enter a denser medium, such as glass, the differential slowing becomes very evident to us.

Figures 25 and 26 are scale drawings to show the truly small degree of refraction differences. These diagrams show the cross-section of a very large equilateral crown glass prism. In Figure 25, the angle of incidence of

CROWN GLASS

	Blue	Red
Index of Refraction	1.528	1.513

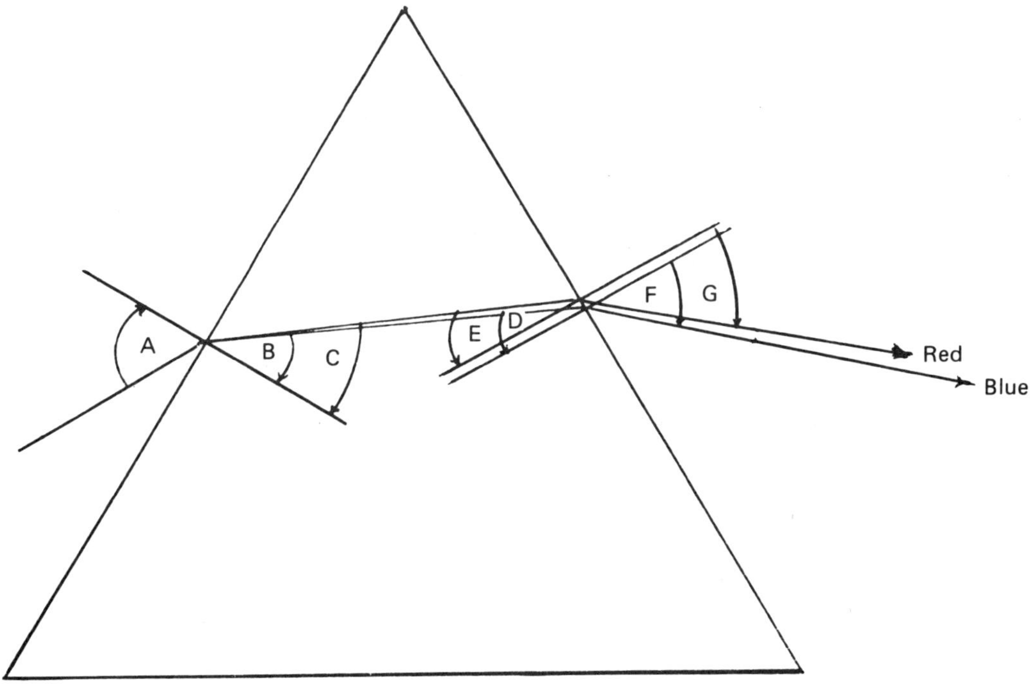

Figure 25. Scale drawing of refraction of white light by crown glass prism.

A = Angle of incidence = 60°
B = Angle of refraction, blue = 34.525°
C = Angle of refraction, red = 34.917°
D = Angle of incidence, blue, in glass = 25.475°
E = Angle of incidence, red, in glass = 25.083°
F = Angle of refraction, glass to air, blue = 41.088°
G = Angle of refraction, glass to air, red = 39.897°

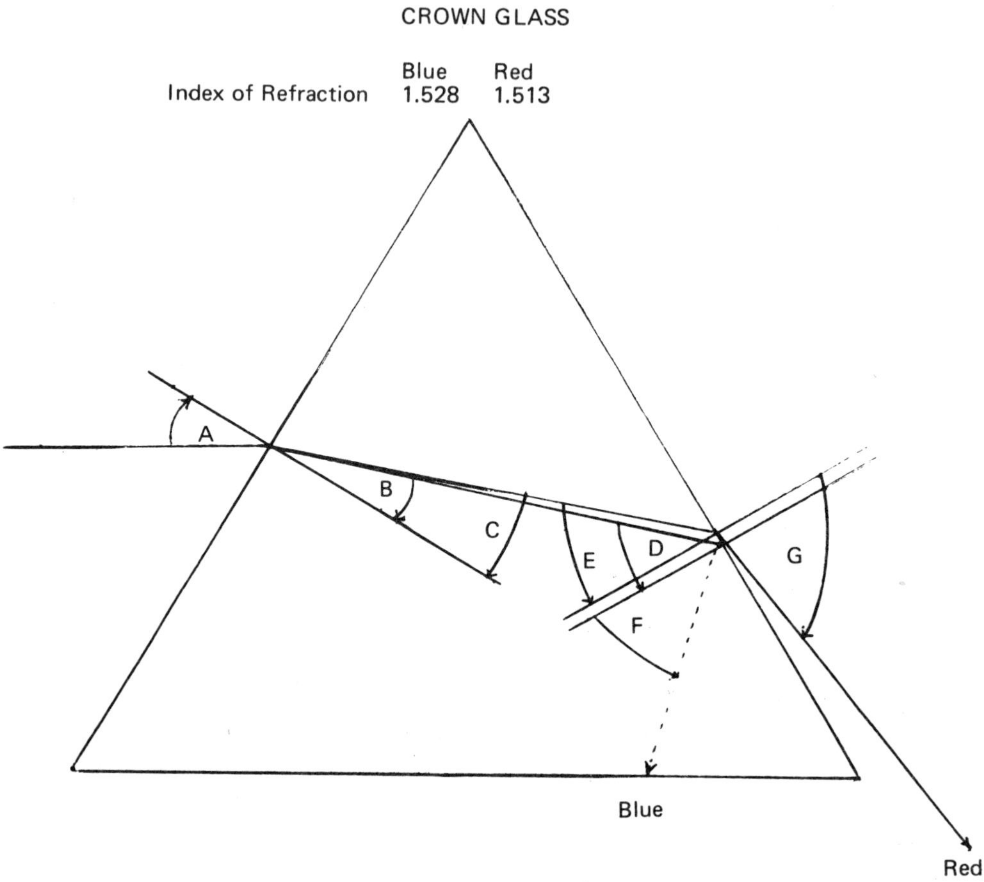

Figure 26. Scale drawing of refraction of white light by crown glass prism, with total internal reflection of blue light component

A = Angle of incidence = 30°
B = Angle of refraction, blue = 19.100°
C = Angle of refraction, red = 19.297°
D = Angle of incidence, blue, in glass = 40.900°
E = Angle of incidence, red, in glass = 40.703°
F = Angle of internal reflection, blue, in glass = 40.900°
G = Angle of refraction, red, glass to air = 80.878°

Critical angle for blue is 40.878°
Critical angle for red is 41.371°

white light is 60°, and it is seen that the difference in the angles of refraction of blue and red light is only 0.392°. In leaving the prism, the difference in the angles of refraction of blue and red light is greater, 1.191°. Figure 26 shows the angle of incidence of white light to be 30°. The difference in the angles of refraction of blue and red light in the prism is 0.197°. In this case, since the angle of refraction of blue light, on exiting the prism, would be greater than the critical angle, the blue light is totally reflected, internally, and only the red light exits the prism.

We have suggested that light of different colors travels at slightly different velocities in the emag sea. Our interpretation of the phenomenon of light refraction, according to the emag hypothesis, suggests that light of different frequencies is slowed in a more refractive medium, as the frequency. As pointed out earlier, the refraction of light varies, in some manner, inversely as the *angle of refraction.* We have shown that the indices of refraction of red and blue light are different in crown glass, and we contend that the velocities of blue and red light vary inversely as the indices of refraction. Since the indices of refraction vary as the frequencies, the velocities vary inversely as the indices of refraction.

$$n \propto f$$
$$V \propto f^{-1}$$
$$n \propto V^{-1}$$
$$V \propto \mathrm{n}^{-1}$$

The velocity of white light in the emag sea is c: 2.9979246×10^8 m/sec. If we divide this by the index of refraction of dry air at Standard Conditions of Temperature and Pressure (STP), 1.0002926, we find the velocity of white light in air to be 2.9970476×10^8m/sec.

$$\frac{\text{Velocity of white light in air}}{\text{Velocity white light in emag sea}} = \frac{n \text{ in emag sea}}{n \text{ in air}}$$

Since, by definition, the index of refraction in the emag sea is 1,

$$\text{V white light in air} = \frac{2.9979246 \times 10^8}{1.0002926} = 2.9970476 \times 10^8 \text{ m/sec}$$

Similarly,

$$\frac{V \text{ blue light in air}}{V \text{ white light in air}} = \frac{n \text{ white light in air}}{n \text{ blue light in air}}$$

$$V_B \text{ in air} = \frac{(V_W \text{ in air})(n_W \text{ in air})}{n_B \text{ in air}}$$

$$= \frac{V_W \text{ in emag sea}}{n_B \text{ in air}} \times \frac{n_W \text{ in air}}{n_B \text{ in air}}$$

$$= \frac{2.9979246 \times 10^8}{1.0002926} \times \frac{1.0002926}{1.0002827}$$

$= 2.9970773 \times 10^8$ m/sec.; greater than V_W in air.

$$V_R \text{ in air} = \frac{(V_W \text{ in air})(n_W \text{ in air})}{n_R \text{ in air}}$$

$$= \frac{V_W \text{ in emag sea}}{n_R \text{ in air}} \times \frac{n_W \text{ in air}}{n_R \text{ in air}}$$

$$= \frac{2.9979246 \times 10^8}{1.0002926} \times \frac{1.0002926}{1.0002770}$$

$= 2.9970944 \times 10^8$ m/sec.; greater than V_W and greater than V_B in air

$$\frac{V_B \text{ in glass}}{V_B \text{ in air}} = \frac{n_B \text{ in air}}{n_B \text{ in glass}}$$

$$V_B \text{ in glass} = \frac{(V_B \text{ in air})(n_B \text{ air})}{n_B \text{ glass}}$$

$$= \frac{(2.9970773 \times 10^8) \times (1.0002827)}{1.528}$$

$= 1.962 \times 10^8$ m/sec.

$$\frac{V_R \text{ in glass}}{V_R \text{ in air}} = \frac{n_R \text{ air}}{n_R \text{ glass}}$$

$$V_R \text{ in glass} = \frac{(V_R \text{ in air})(n_R \text{ air})}{n_R \text{ glass}}$$

$$= \frac{(2.9970944 \times 10^8)(1.0002770)}{1.515}$$

$= 1.981 \times 10^8$m/sec.; greater than V_B in glass

To summarize:

	Blue light (4000 Å)	Red light (6000 Å)
n in crown glass	1.528	1.513
V in crown glass	1.962×10^8m/sec	1.981×10^8m/sec
n in air	1.0002827	1.0002770
V in air	2.9970773×10^8	2.9970944×10^8
n in emag sea	1	1
V in emag sea	2.9979246×10^8*	2.9979246×10^8*

*By definition of present usage, V_W, V_B, and V_R are constant.

22. *Summary*

We are now ready for our conclusions. What are the basic principles we have discussed?

1. We have introduced the *small bang* theory to account for the origin of the universe by showing that something can come from nothing if we take into consideration negative mass. Such a notion seems far superior to the *big bang* or *steady state* theories of cosmology, both of which start with *something*. The *small bang* theory starts with nothing, and requires as a postulate, the recognition that *there was a beginning* (postulate 1). There is no need for the irrational concept of first cause, either as a theological notion or as a primordial mass of matter whose big bang resulted in the universe.
2. We have reaffirmed the definitions of matter and mass (postulates 2 and 3) and abandoned the irrational notion of disembodied energy (postulates 4 and 5); and we have postulated that *all action is direct.* All action is the result of bodies in motion, colliding with one another.
3. There is no such thing as the magical notion of one body's influencing another body without an intervening, *real* medium (postulate 7).
4. Charged bodies, therefore, are inadmissible.

Consequences of the above are:

1. All in the universe is particulate.
2. All so-called electromagnetic radiation is particulate.
3. The ultimate particles of matter (i.e., the building blocks of the universe) are the emag of positive mass and the *other emag* of negative mass.
4. All matter in the universe is a combination of emags and *other emags.*
5. The sum total of all is zero.

6. Arbitrarily, we designate *our* matter as having positive mass. Since the mass of *our matter* is, essentially, the mass of the nuclei, protons are of positive mass.
7. Electrons are of negative mass.
8. *Other matter* and *anti-matter* become synonymous.
9. Our new convention permits us to speak of electrons as negative mass, rather than negative charge, and of protons and positrons as positive mass, rather than positive charge.
10. Since most of matter is nuclear, of positive mass, the *emag sea,* or ether, if you will, is primarily of negative mass.
11. It is the negative-massed emag sea that pushes negative-massed electrons around protons, and the planets around the sun.
12. The null result of the Michelson-Morley experiment is evidence for the statement that the satellites of a planetary system ride, passively, in the sea of emags.
13. Consequently, we cannot support Einstein's second postulate of the constancy and the ultimacy of the velocity of light.
14. Lorentz's constant $(1/(1\text{-}v^2/c^2)^{1/2})$ is, as Lorentz himself said, a useful mathematical device to make Maxwell's equations applicable in any frame of reference.
15. The Bohr "stationary state," unexplainable by previous methods, corresponds to the null result of the Michelson-Morley experiment on the atomic scale. It represents the passive riding of the hydrogen electron, in the emag sea, around the proton.
16. Consequently, applying Kepler's third law to any planetary system, it is possible to determine the passive frequency of the hydrogen electron around the proton. It appears to be 0.01000×10^{15}/sec.
17. Using conservation of momentum, it can be shown that the "photon" has mass, and that mass varies as the frequency of emission.
18. The "photon" is considered to be a group of emags, *other emags,* emag aggregates, and *other emag* aggregates rather than a discrete particle.
19. Since current thought considers the neutron to be of zero charge, and since the concept of charge is unacceptable in the emag hypothesis, the notion of the neutron as a discrete particle is abandoned, and is considered, once again, as Bothe and Becker did originally, to be a "beam of gamma rays."
20. Wave phenomena are interpreted as the effects of hordes of particles—such as a wave of water, which is known to be composed of hordes of water particles.
21. Quantum theory is accepted as a useful, statistical analysis of probability.
22. With the abandonment of the concepts of charge and action at a distance, it is shown that the constancy of G in the universe is suspect.

23. So-called binding energies are shown to be due to "holes of zero mass" in nuclei.
24. It is shown that these "holes of zero mass" are related to what we recognize as magnetism.
25. "Field" equations are not challenged, since they work, but "field" concepts are reinterpreted as the effects of emag-*other emag* interactions; distant interactions being equated with so-called gravity; near effects being equated to so-called electricity; intermediate effects being equated to so-called magnetism; and very near effects being equated to intranuclear forces.
26. Chemical reactions are described in the light of the emag hypothesis, and the notion of "shared" electrons in the present-day sense is abandoned in favor of electron orbital linkage.
27. Batteries and electric currents are described as effects of *other emag* motion, rather than electron motion.
28. The phenomena of diffraction and refraction are described as particulate phenomena.

The thesis has been presented. It would be too much to expect that it is beyond challenge, or that it will not be challenged. It would be too much to expect that the thesis can withstand all such assaults. It is not too much to expect that there is much of value in this thesis. It should be challenged. Progress can only be made by proposing a theory, challenging it, modifying it, and testing it.

Index